Practical Solutions for Energy Savings:

A Guidebook for the Manufacturer

Practical Solutions for
Energy Savings:
A Guidebook for the Manufacturer

Roger Brown

Routledge
Taylor & Francis Group

LONDON AND NEW YORK

Published 2020 by River Publishers
River Publishers
Alsbjergvej 10, 9260 Gistrup, Denmark
www.riverpublishers.com

Distributed exclusively by Routledge
4 Park Square, Milton Park, Abingdon, Oxon OX14 4RN
605 Third Avenue, New York, NY 10017, USA

Library of Congress Cataloging-in-Publication Data

Names: Brown, Roger, 1958- author.
Title: Practical solutions for energy savings : a guidebook for the
  manufacturer / Roger Brown.
Description: Fairmont Press, Inc. : Lilburn, GA, 2018. | Includes index. |
  Identifiers: LCCN 2018015674 (print) | LCCN 2018028454 (ebook) | ISBN
  9788770222600 (Electronic) | ISBN 0881737887 (Electronic) | ISBN
  9781138311329 (Taylor & Francis distribution : alk. paper) | ISBN
  0881737879 (alk. paper)
Subjects: LCSH: Industries--Energy conservation. | Industries--Energy
  consumption.
Classification: LCC TJ163.3 (ebook) | LCC TJ163.3 .B77 2018 (print) | DDC
  658.2/6--dc23
LC record available at https://lccn.loc.gov/2018015674

Practical Solutions for Energy Savings: *A Guidebook for the Manufacturer/Roger
Brown*
First published by Fairmont Press in 2018.

Routledge is an imprint of the Taylor & Francis Group, an informa business

10:0881737879 (The Fairmont Press, Inc.)
13:9781138311329 (print)
13:9788770222600 (online)
13:9781003151319 (ebook master)

While every effort is made to provide dependable information, the publisher, authors, and
editors cannot be held responsible for any errors or omissions.

The views expressed herein do not necessarily reflect those of the publisher.

Table of Contents

PART I—What's in This for You? 1
Chapter 1
How to Lower Your Costs Fast 3
    The Main Problem. 3
    Chapter Summary. 20

Chapter 2
Three Case Studies—Can You Really Save Millions? 21
    Case 1: Multiple large pumps in Texas 21
    Case 2: Paper mill in South Carolina 23
    Case 3: Small operation that saved 40% 24
    The Bad Thing about Case Studies 26
    What Have We Learned from the Case Studies?. 29
    Chapter Summary. 31

PART II—Tools for Demand and Cost Reduction 33
Chapter 3
Demand Response/Interruptible Service 35
    The Origins of Interruptibility 36
    Details You Need to Know .37
    Why Utilities Have Interruptible Service. 38
    Also, You Need to Meet the I.S. Program Manager 42
    An Example of Interruptible Service That Works 43
    Economic Curtailments. 43
    Chapter Summary. 46

Chapter 4
Thoroughly Understand Your Relative Risk to Win with Interruptible Opportunities 47
    In Deregulated Power Markets, Things Work Like This... . . .47
    So, What Options Are Available to Reduce Costs? 48
    Three Keys to Understanding Bills 52
    Why Would a Utility Care *When* You Use Your Energy? 53
    Understanding How Your Utility Buys Power. 54
    Reliability Is in The Eye of The Beholder. 62
    Chapter Summary. 66

Chapter 5
Real Time Pricing (RTP). **67**
    More on Real-time Pricing 68
    Chapter Summary. 71

Chapter 6
Coincident Peak Costs. **73**
    What Is It? 73
    How It Works 73
    Why Coincident Peak Programs
        Can Be So Powerful for You 76
    Chapter Summary. 77

Chapter 7
Solar Power **79**
    What Does It All Mean for Manufacturing Plants? 79
    What is Solar Exactly, and How Does It Work? 79
    Solar Roof from Tesla. 87
    Net Metering and Renewable Energy Certificates. 89
    How to Compare Your Solar Storage Options 92
    Chapter Summary. 95

Chapter 8
Energy Management Systems **97**
    Components of an Energy Management System (EMS) 97
    Sources of Funding for Energy Projects. 103
    Chapter Summary. 105

PART III—Free Money 107
Chapter 9
Sales Tax 109
    Sales Tax Exemptions. 109
    Which Businesses Are Exempt and Why? 110
    Refunds for Overpayment 111
    States Which Allow Manufacturers to
        Exempt Sales Tax on Utility Bills 112
    Chapter Summary. 116

Chapter 10
Billing Line Items You Can Remove **117**
    Ratcheted Demand Charges 117
    Systems Rentals/Facilities Costs. 118
    Chapter Summary. 122

Chapter 11
Combining Meters **123**
    Why So Many Meters? 123
    Options You May Have for Meter Combination. 125
    Quick Primer on Power Delivery—
        Centralized Power Generation. 127
    Chapter Summary. 131

PART IV—How to Get Things Done with Your Utility **133**
Chapter 12
Dealing with Your Utility. **135**
    How a Utility Works 135
    How to Work with a Utility Most Effectively 142
    Chapter Summary. 145

Chapter 13
Appoint a Champion and Get Things Done. **147**
    Why People Have Trouble Implementing Things 147
    Who Should You Choose to Be Your Champion? 151
    Projects Managed the Navy SEAL Way. 152
    The True Value of Third-party Oversight. 153
    Chapter Summary. 157

Chapter 14
Long-term Monitoring and **159**
Acting on Your Data **159**
    How Do You Accomplish Big Things? 160
    Action Plans 162
    Chapter Summary. 163

Chapter 15
Summary and Plan Going Forward **165**

The Challenges Ahead 165
Key Points of the New Approach 165
Ancillary Benefits of Energy Saving. 167
Final Thoughts. 169

Addendum. **171**
How Having DSIRE Can Save You a Lot of Money. 171
Summary 172
Mindset. 173
Five Energy Trends 179
Inertia of the Status Quo 184
CO_2 Science and the Battle to Save the Planet 185
Why Excess CO_2 Makes Temperatures Rise 186
Is Energy Creation the Only Cause of Atmospheric CO_2? . . 188
What Happens If We Do Nothing? 188
Global Emissions by Economic Sector 189
Countries that Contribute to CO_2 Emissions. 190
What Can We Do About It? 190

Index **195**

Introduction

I have been an energy consultant for manufacturers for the last 25 years, working in every U.S. state and Canada, saving my clients millions of dollars.

In this book, we will get right to the point. I will let you in on my cost reduction secrets and how to get them done. That is, *how to slash your energy costs without investing big money.*

We will talk specifically about my three pillars of cost reduction:

1. Assemble your options and analyze your relative risk—your current exposure and the exposure that your cost-saving opportunities present to your business.

2. Think win-win with your utility—to develop your options, first understand what drives *their* costs and then help them drive them down. The savings will be passed along to you.

3. Then, perpetually cut out obvious waste in your operation.

    Thus, my goals for this book are simple:

*   Show you how to cut significant energy costs at your facility without spending much by managing demand and other billing factors.

*   Motivate you to act.

Manufacturing leaders need to have confidence in the actions they can take to cut their costs. While many of the ways to reduce costs are simple, the road to full implementation of those ideas is often a difficult and long one. There are "we can't do that" forces within our organizations that are intent on silencing best efforts to make any changes.

It will be no surprise to you that I have found manufacturing plant employees often like to keep the status quo intact. These are the kinds of things in play in our plants that will quickly rise and conspire to kill projects like the ones I will be suggesting you launch—even projects with enormous cost savings and quick paybacks.

I have seen too many great projects sit on shelves or get shouted down to protect turf and internal fiefdoms. Even armed with significant cost savings ideas, your best efforts will fall flat if you don't plan how to circumnavigate these inherent internal obstacles.

Is this book for you?

I wrote this book for an exceptional class of utility customer—leaders of manufacturing businesses charged with reducing costs. If you are such a leader, you may be familiar with many of these concepts. As busy leaders, you often do not have the time to drill down to the level of detail required to understand these options and make them happen.

These are the people and job functions that will typically benefit most from learning the information in this book:

Business Owner

Because an owner stands to benefit the most directly from increased profitability, they also have the most to gain by focusing on energy savings. The business owner alone has the most incentive to save money and to spend as little as possible to achieve those savings. The business owner often must make personal sacrifices if costs are not reduced. In competitive markets, energy cost savings can serve as a shield against rising raw materials costs and shrinking retail prices that can rob a business of profits.

CFO

The CFO is the person responsible for being the cost-cutting metronome for many organizations. The CFO has a significant advantage over many other job functions in this regard since they oversee *all* costs and do not tend to be territorial and biased in favor of focusing on one area. The net dollars do the talking. They don't care which cost center savings come from.

Plant Manager

Energy probably is or should be a critical topic of discussion in weekly meetings. If the plant manager doesn't make energy a visible priority, the people on the floor driving the bill won't take their role in cost reduction seriously. Energy can be a pivotal cost category to drive success in other areas of cost reduction. Why?

Because the wins can be huge and the process of cost reduction in utilities can serve as a template for getting things done in other cost areas in the organization.

Plant Engineer

Much of what a plant engineer does goes unnoticed because the work is seen as maintenance. So, plant engineers have an opportunity with energy cost reduction to impact the organization and significantly distinguish themselves by recommending dynamic changes to reduce energy costs.

Energy Manager

Anyone tasked with reducing energy costs at a manufacturing facility will be able to use the concepts in this book to quickly lower costs in most states by anywhere from 10-25%.

Overall, this book is a handbook for all those tasked with taking costs out of the energy equation. Having cost-saving ideas is the easy part. Getting them done is more complicated. It is easy for projects to get shut down summarily by others who fear change. There are always a lot of sensible sounding reasons not to act and stay where you are as an organization.

This is especially true in the energy space. That is because many of the savings strategies are not visible and require a thorough understanding of the risk-reward potential. As you will find, that potential takes work and a good bit of time to uncover. This process of discovery shows the power of persistence and the ability to probe for detailed answers.

Use This Book as Both a Guide and Call to Action

If you fear taking small risks, achieving energy savings will be difficult for you. The cost reduction offerings require you to analyze a lot of elements of energy supply at a detailed level—probably in a lot more detail than you usually go into.

Your local utility rep is not going to try to talk you into doing any of these things either or give you much assurance you are on the right track. It costs the utility money to deliver savings to you.

And, since it may increase the risk for you, the customer, your utility will be more than happy to let the fear of the *perceived* risk talk you out of moving forward.

But this struggle is very worth while for you and your company. There is a massive opportunity in energy for your business to save a lot of money. It is also an opportunity for you to stretch the boundaries of your comfort zone and accomplish something that will pay significant dividends for years to come.

It is not tilting at windmills to focus a bit on the "save the planet" aspect of reducing your energy costs either. Cutting back on what your factory uses will reduce a world-sized problem.

It is often difficult to see the role that we as individuals have in the big picture. Many others in faraway lands are involved in making larger-scale changes that increase or decrease the amount of atmospheric carbon dioxide by a lot more than we can. It is too easy to write the issues off as hopelessly political and take sides that our favorite red or blue politicians take.

No matter which job group you find yourself in, if your job touches energy costs for your company, this book will help you get results faster and get solutions with a lot more confidence. Having confidence may seem too rah-rah for the sedate business world but believe me, it is essential to the eventual positive outcome.

I will go so far as to say, energy should become an obsession for somebody at your company. The only reason it isn't now is that the usage numbers are hidden, locked away in the utility's meter that is located somewhere behind one of your buildings.

What You Should Expect from This Book

We are going to distill what savings are possible and how you can quickly accomplish those savings. You should expect to walk away at the end of this book with confidence and a realistic plan of action for reducing your costs. I have found that the primary reason why people are not successful at reducing their costs is they do not believe it is even possible. It takes willpower and determination to stick with something when you have not seen others do it first.

It is much more comfortable to stick to the job you were "hired to do" and forget about the energy savings projects we will be talking about that seem to be extracurricular.

I say extracurricular because, unless you are an energy manager for a living, you probably have a primary job function that takes up most of your day. In that case, anything we will discuss will lie outside of that job description. You will be picking up the energy cost savings baton because you know it is the right thing to do. I want to prepare you for the long-term battle ahead.

Although many of the ways we will discuss to save are very simple, we will be spending time discussing the getting-things-done part. You and I both have seen too many great projects die on the vine because of internal politics. Being honest with where you stand in knowing what your options are for dealing with the internal forces of "inclusion" and "consensus management" is as important as knowing what the savings methods are themselves.

Companies always have options for savings. The savings you find will create momentum to find other cost areas from which to exorcise waste demons.

There is a lot of fear inside manufacturing operations that brings on this internal project-killing stress. Fear of change. Fear of doing something or anything different. Maintenance departments are often tasked with energy oversight and are run by territorial and controlling personalities. I want to disabuse you of the notion that if something is the "right thing to do," it will somehow happen. It won't. Getting things done almost always involves going around naysayers.

I have been down these savings pathways for years, and I know how to get things done in a plant environment. If you follow the steps I lay out, you will be well equipped to steer clear of the naysayers that will try to crash your party.

You can save money with utilities without having to spend any upfront capital. Believe it or not, that is where the most significant savings are.

Part I

What's in This for You?

Chapter 1

How to Lower Your Costs Fast

THE MAIN PROBLEM

A Word about Savings and What is Possible

Understanding what others have done and developing your belief that you can do it too is the most important thing I'm going to show you. Let's get started right off the bat by focusing on what you can do to set a very short initial timeframe for getting things moving. There is no reason why you can't achieve your savings goals within a few months.

Getting things done *efficiently* is the challenge. You have a lot of other things on your plate. We all do. These projects will have to be worked-in among a host of other commitments and deadlines. It is very easy to procrastinate with work that takes a long time and to drop some things altogether in favor of more rewarding short-term to-do items. Once you realize the exact tasks you need to get done, it becomes easier to allocate the time to do them.

I wrote the book to break down tasks and make them easier for you. I have been through this exercise hundreds of times, and I know how demoralizing it can be to work on projects where the feedback is infrequent and you wonder if you are doing things correctly. Once you have seen how successful this process has been, it will be easier for you to follow through with your own efforts at cost reduction.

It is painful to start long-term projects that seem to have no end, especially projects where the financial upside is not clear. I know that, and I want to encourage you to stick with it. We are going to discuss strategies that can mean the difference between success and failure as a business. That is how large some of the financial rewards for these efforts can be. It is not uncommon to see a 10-15% cut of energy budgets with the strategies we are going to discuss.

We are going to talk mainly about electricity and electric utilities. So, when I refer to "utilities," I will mean electric power providers. We can apply the same principles of cost reduction to all the other areas that fall under the utility rubric, but it is best to focus on one utility at a time, because success in one area will give you the confidence and drive to apply what you learn to other areas as well.

Why focus on electricity? Because electricity is where the most significant money will be for you; it always is. If you concentrate on electricity first, you will save the most money in the shortest time. And, the general principles you apply in reducing electricity costs will be effective for all the other utility genres as well.

There are a lot of dollars now hidden in your utility invoices regardless of the size of your company. I have seen savings happen in big plants and small plants, with those companies spending almost nothing to achieve savings. The concepts we will discuss are not targeted at only substantial accounts. I have seen relatively small utility customers cut their bills in half by using these strategies.

This reduction does not just occur in some states, or with some utilities, as many would have you believe. I have seen manufacturers achieve significant savings in almost every U.S. state and Canadian province over my 25 years in the business.

I had heard it said that utilities have tightened up their act and they just don't make the mistakes they used to back in the day when they hand-calculated invoices for you. Not true. While it may be true that good old-fashioned math errors may be less frequent, there are still lots of ways utilities make mistakes.

Public Perception of Utilities and Their Prices

Once upon a time, all utilities, like airlines, were regulated. That meant the prices for the energy you saw posted were the ones you paid and that was that.

Customers thought you just got your utility bill in the mail and you paid it. There was nothing more to talk about. "Cost reduction" meant paying the darned thing on time to avoid the late charge.

That is not the case now—that does mean you will not encounter a fixed mindset when you deal with your local utility, especially at the customer service-level of these organizations. Utilities attract a very loyal type of employee, and their attitudes are old school and die-hard.

The reason for the attitude stems from the fact that local governments had initially set up all local utilities as monopoly suppliers. If you lived in a geographical area, you were forced to buy your power from them. From the perspective of the utility, they did not have to try to win your business or try that hard to make you happy to make their business model work.

Utilities are still regulated but a lot less so than they used to be. And that transition from controlled to less controlled has helped our efforts to cut costs tremendously by opening their thinking.

How Electricity Deregulation Came About

"Deregulation" entered the scene in 1999, with California being the first state to jump in with both feet. As of this writing, twenty-two states have since deregulated. However, deregulation has not changed the world the way the originators intended. It has also not made everyone blissfully *liassez-faire* happy like we thought it would back in 1999. It did, however, change many utility mindsets and made them more customer service-oriented. This change in mindset has opened options that can reduce costs for manufacturers as you will soon see.

The bottom line: larger public-owned utilities have developed more of an entrepreneurial way of doing business. People went to work at utility companies who thought more like business people. Utilities came to realize that their game of monopoly would be over soon and that they needed to start behaving a lot more like a business in a free market economy.

They Still Don't Mind Telling You "NO"

Utilities are very stable entities to work for, which is essential to understand when dealing with them because of the type of employee they tend to attract. Seldom do they lay people off. Rarely do utilities reduce salaries to compensate in bad times. (They just rebill their customers at higher rates if their costs increase.) Most people who go to work for utilities thrive on stability, not on pushing the envelope.

You have probably found that at the customer service level, utility behavior resembles that of your local government tag office. "You are lucky to have us here, so sit down, take a number, and don't be in such a rush!" Often, it seems a bit like they want you to go away so they can get back to moving papers around.

This does not mean that everyone at a utility thinks or behaves this way. Far from it, but you may have to work at it to find the thinkers. While saying "NO" as a knee-jerk reaction may be the mindset of many, it is far from being the only mindset there is. I have usually been able to find someone at every utility that refuses to follow the bureaucratic pattern and who wants to help.

Unfortunately, that person does not have a standard title, so they are not easy to find on an organizational chart or in the phone directory. This person may be your local rep, but large utilities usually do not assign a local rep to accounts unless you are a giant company spending $1,000,000 or more a year with them. It could be someone in the business development end of the utility. These people are involved in winning new business, so they must be out-of-the-box thinkers.

If you become frustrated getting to the right person at your utility and just can't find anyone who cares, another strategy is to reach out to your local chamber of commerce or state economic development agency and ask for their utility contact. The person interacting with the community will have been chosen because of their big-picture mindset toward business.

When You Get to the Right People,
Utilities Can Become Downright Entrepreneurial

If you are the type of leader who does not understand the word "no," that is *good*—if that is not the case, I encourage you to learn to become that type of leader. If you want real results, working with utilities will often require you to hear the word "no" or something similarly discouraging to your plans and efforts, multiple times.

I want to encourage you that there is a lot of savings waiting for those who persevere and push past this attitude. The solutions can come on the one-yard line.

Politely asking multiple people the same question can often yield better answers. I never consider a "no" to be a "no" until I have checked with numerous people, especially if answers are confusing and are not in the best interests of those involved.

The higher you go in the organization, the more flexibility the person will have to handle your question. That person can either answer your question or get you to the right person. But, in either case, aiming high up on the organizational chart yields the most efficient use of your time.

Having "polite persistance" is the master key to getting things done in the utility world. Never accept an answer that seems short, regardless of how emphatically it is stated or by whom. Adopting this philosophy will put you at the head of the line and get answers more than any other suggestion I can offer.

The Main Thing to Remember—
Utilities Are Not Obliged to Tell You How to Save Money

None of us ever expects the IRS to tell us ways to save on our taxes, but we somehow believe our utilities will show us how to reduce their invoices to us. While there are potentially more ways to save money now than ever before, do not expect your utility to offer them up to you on a silver platter. But, don't get mad at your utility for not telling you these things proactively. At the end of the day, they are under no obligation to help you save money. You must initiate and follow up on the savings process. The utility that serves you is not in the business of having their representative teach customers how to reduce utility revenue. Think about it, would your own business be likely to hire someone whose job it was to reduce billings to customers? Probably not.

While you may have received a courtesy "heads up" about a new rate program or service, your utility is not likely to take you by the hand and make you do it. They have let you know about it and nine times out of ten that is all they are going to do.

The utility representative who gave you the "special update" may also discourage you from taking advantage of it. We will go into more details about this attitude a bit later and how to mentally process it. For now, the burden of the job of cost reduction falls squarely on your shoulders. Realizing this will reduce your stress level and give you a mindset that will translate into more success.

Understanding and dealing with the way utilities think will be a significant factor for you in getting your costs down. Why? Because many at your company will blindly trust utility employees to be looking out for your best interests. You might have to fight that intra-company perception as well when you hear successive negatives.

I am repeating myself, but it is critical you wrap your mind around what I am saying. Understanding and dealing with the mindsets you will encounter at utilities and at your company is essential to a successful project. It doesn't matter how significant the potential savings are or how small if a project is perceived as more risk than reward it will be impossible to get it implemented. Perception is everything of course.

The negative pattern of thinking we have been discussing is not just present at electric and gas utilities, it is a virtual certainty, that you will encounter a version of this territorial and monopolistic thinking at your own company. I ran into a situation once where senior level executives at a company were working with senior level utility managers in the community, and the company manager would not implement obvious savings for fear of offending the utility manager. Hard to believe, but true. And keep in mind these voices of unreason could be very loud, and the people broadcasting them can seem very sure of themselves.

"No" As the Default Answer

I have been working with manufacturers for my entire 25-year career. I must say, nothing much has changed about the smoke screen many working inside those organizations throw up at the sign of any change.

While the names are different, the reactions are similar—people who don't understand the details thoroughly want to protect their turf. They are only comfortable dwelling in the land of the status quo.

This resistance is not only occurring within manufacturing. Resistance to change is a universal human trait and will shut down even the most obvious organizational improvements.

Savings Will Be Significant but Getting to it Can Be Confusing

The options that we must use to reduce costs significantly without spending money are poorly understood by most. When utilities talk about their latest rate plan to their customers, it is usually accompanied by a confusing mélange of terms. Doctors with their befuddling lexicon are easy to understand when you compare their insider talk with that of your favorite utility. Try reading one of your utility's rates. Terminology and utility-speak are often used to obscure rather than clarify.

Our Own Motivation is Often Unclear However,
and Motivation Determines Drive to Succeed

At times, we all lack the strength to follow through on our plans to do worthwhile things. That is because our motivations to act are weak and unclear. Often, we just do not really know what is possible and *when we don't know, we shoot low*.

Maybe you have tried to make savings happen before at your company and have gotten nowhere. Company stories of big project flops and how things "don't work that way around here, " will subvert the efforts of even the most stalwart.

The broader value to the organization of any savings project is hard to nail down without a heck of a lot of effort. Therefore, it is not only essential to understand the rewards specific to the project but also its ancillary benefit. That means realizing not only the hard dollar savings, but also the soft dollar savings. Momentum and the contagious nature of success are vital factors that influence hard and soft costs. Do not ignore soft costs just because they are difficult to quantify.

If you focus solely on trying to prevent the worst-case scenario you can talk yourself out of taking almost any action in life. It is important to compare the relative risk that those options present. The problem in evaluating the full value of our alternatives is that we don't understand our present position well enough to accurately contrast it with the alternatives.

We often incorrectly begin our analysis of options under the assumption that our current situation bears no risk at all. The truth is that *all* scenarios contain inherent risk, including where you find yourself now. Your current situation may be presenting more risk than the alternatives. You just don't "feel" current risk anymore because you have adjusted to it over time and are, therefore, not aware of it.

On top of the stories of project failure from inside your company and how things aren't done that way around here, your utility will not work to make things any clearer. Utilities, especially, don't like it when the change is not their idea.

People enjoy the false security that inaction can provide. The deer in the headlights syndrome is what we call it when it happens to other people. People feel like if they are paying their bills and no wild beasts are chasing them, then everything is ok and they don't need to act to improve.

Lack of Deep Understanding Can Kill Projects

Old school energy managers and others who find themselves thrown into a position of overseeing energy in industrial plants have a lonely plight. That is because the most that many folks have thought about energy costs is that you must pay the bill for your power or your energy will get cut off.

Face it. Most things in life just do not take that much time to analyze and act on. With the many decisions we make every day, it is no surprise that all of us come up with off-the-cuff shortcuts for making decisions that get us through but don't consider all the relevant variables. Shortcuts are a necessary, in fact, or we would spend all our time trying to figure out what to have for dinner.

A problem arises, though, when we handle more complex issues via the same shortcut decision-making process. It is very seductive to make all decisions using mental shortcuts. Since everyone else is doing this too, we don't get much pushback. When we do this with more complex problems, we sub-optimize and choose poorly.

With complex situations that we barely understand, this type of cursory thinking will feel just as meaningful as thorough risk analysis. That is because everyone around us is thinking the same way—by the seat of his or her pants. We are left with a culture of shoot from the hip decision-making that continually has us bringing a knife to a mental gunfight.

We especially run into problems with energy decisions in this regard because everyone thinks they know a little about energy. And when everyone thinks they know, it becomes challenging to detect mistakes. Through our discussions in this book, I am going to let you in on the truth about what your options are and the level of depth of relative risk analysis that needs to be insisted on to make the best decisions.

Concept of Relative Risk is Also Not Understood

The point I want to make is this: You are going to have to get approximately ten times more information than you typically do to accurately analyze the risk in energy cost reduction scenarios. The concept of risk analysis and interpretation is a topic that we will continue to discuss.

The nature of energy cost savings is different than other industrial categories of savings in this way. These projects simply do not fit neatly into a traditional ROI analysis like buying a new press brake because implementation cost is meager and the savings going forward can be very high.

It Is Easy to Give Up

When the waters are full of sharks, it is easy to rationalize jumping out and walking away. Typically, there will be no one within your organization that will question you if you say, "this project just adds too much additional risk. I recommend we not pursue it." Your company may be missing out on a fortune in cost savings, but no one will ever realize it because there is typically no post-game analysis.

The only reason consultants like me can put food on the table is manufacturers and processors have internal impediments to analyzing risk and making decisions.

The Local Utility Model Doesn't Help

Your utility has a vested interest in having aggressive power rate formula options *on paper*, but most of the representatives the utilities have working with you would never in a million years do anything to add perceived risk. That makes it very difficult for them to recommend their programs to you.

Utilities are a "cost plus" business. This means utilities add up their costs, add on a percentage for profit, divide that by kilowatt-hours, and then bill you to recover what they have spent. They can be as sloppy as they want in making energy and they still win at the end of the day.

Does any of this sound like a system you should trust to help you lower costs?

But All Is Not Lost

Take heart. I understand your plight. I have seen many companies save millions wrestling these same alligators and you can too.

I am spending time on the psychological side of cost reduction because I have seen just as many companies fail to achieve results for interpersonal or motivational reasons. The internal company storms derail many efforts. Cultures are, of course, created within companies in which people must get along since they spend most of their waking hours with these other people. Maintaining these relationships and not wanting to challenge the strongly voiced negative point of view knock many great projects off the rails.

This negativity is why it is often easier to get things done when something culturally disruptive happens, like when one company is purchased by another or a new CFO joins the team. There is no cultural debt of silence or inaction to be paid as a sacrifice to maintaining the peace. Simply said, project leaders are free to do the right thing without paying much of a social penalty.

One Thing We All Know—Energy Costs Are Going Higher

Manufacturers have increasingly high energy costs, and these internal and external interpersonal battles are not optional if we want to lower costs.

But, how do we know prices are going up with such certainty? Simple: the fuel that generates electricity is "fossil" fuel and is therefore in finite supply since it takes millions of years for carbon life forms to become oil and coal. Our supplies of both of those raw materials will be much more precious one day. No one on the planet knows with any accuracy when that day will come.

I am Betting, You Won't be Surprised by What I Tell You

King Solomon said there is nothing new under the sun. The savings ideas we will discuss are not going to be news to you in many of cases. There are no perpetual motion machines out there. And you are very likely to have heard of or you will be somewhat familiar with the concepts in this book.

I can even imagine the "new program" information your rep brought you is sitting somewhere in your office now on the shelf, acknowledged but not analyzed further or acted upon.

Lack of Ideas is Not the Problem

So, the challenge in saving big money for your operation is not going to be *finding* the ideas themselves. There are loads of them. The problem is sizing them up correctly and taking action to get them done.

Most Solutions Will Not Affect Plant Operations

The good news is most of the projects we will discuss are "paper projects." That means they will take place on the utility bill—in other words, the formulas and details of billing you see in the mercilessly complicated invoices you receive.

My goal is for you not to spend a dime before you receive the results. We do not want to impact the people on the floor or their work if we don't have to.

This is How a Project Often Goes off the Rails

I worked on one recent project that would have delivered approximately $100,000 in annual savings and saw it derailed by internal forces without due process. We will affectionately name this problem the "Maintenance Manager Syndrome."

This gentleman did not want to hear anything about savings. Every question he asked yielded yet another question. He was determined not to be swayed, and he was relentless. The more someone like this articulates their position, the more dogmatic they become. Your cost-saving projects cannot afford to be plagued by this dynamic.

He was the manager in charge of energy for a large hospital, but he could have been any middle-level manager in any large operation. He had been there for over 30 years and oversaw maintenance at the facility. You know him—he thinks everything he came up with is a good idea and everything he didn't think up is a dumb idea.

Why is This Archetype so Negative?

The forces inside the operation that created him are a real problem with American businesses. Ego and territorial markings have become more important to him than profits or the success of the organization to him. He will do anything to protect the status quo. Why? Because there is nothing in it for him to push the envelope.

What causes this behavior? It is the desire to be right at all costs. If it has not been done, there must be a good reason why it wasn't done. It justifies one's existence in the organization to say "no" and protects one's job. Large egos lord over manufacturing plants.

The Costs of This Behavior Are High

The cost to businesses for people being territorial and rejecting beneficial projects is both tangible and intangible. It is tangible in the direct impact of lost savings on projects that never get done. It is intangible, too, with the impact that the lack of getting things done has on others who have ideas and are intimidated into not voicing them.

Dealing with these people should be an essential part of your project troubleshooting and is no small task. Running intra-company interference is every bit as important to your goals as the nuts and bolts of analyzing risk and reward. Why? Because others listen to them and a casual negative comment can feed into others' desire to shelve anything that brings even the hint of added risk.

This Individual is Seldom Challenged by Higher-ups

Of course, these folks are seldom ever challenged by higher-ups. Why? Because they are "needed to run the plant." They are the ones who stay late and fix things and come in on Sundays. The rationale is that you can't make them mad by challenging their attitudes or else they might leave and where would you find another one?

The Bottom Line We Must Recognize

So, unless we strategize around it, manufacturing plants are straight-jacketed by the most conservative no-action taker. If we are going to succeed in reducing costs, we are going to have to acknowledge their presence.

It is ultra-costly to manufacturers, and that is why we must develop workarounds for the "automatic" naysayers.

Making big projects work and saving a lot of money is all about pushing boundaries and being robust in the face of the loud voices dominating manufacturing plant cultures.

Do We Include the Naysayers in the Dialogue or Not?

The naysayers do need to have a seat at the table. There is no way around that. We need to have them see themselves as part of the solution. The more naysayers are involved in data collection for example, the more they take ownership of projects.

The more they feel the decision to take on more risk will not be solely blamed on them, the better. You must be careful with the egos involved here. Ego and reputation protection is a significant element driving people to take actions. Often, people will be on your side in such a debate but speak against your cost-saving project because they have a reputation for fighting against what the engineering department wants to do. They may feel they must speak out against your project because that is just what they do.

You can draw others to your side of the table by forming a group to address the savings opportunity and involving them in it. Making everyone eat supper on the same side of the table helps counterbalance intra-plant conflicts.

As the naysayer gradually feels like their interests in the new group outweigh their other cultural affiliations, the quicker they become an aid and stop being negative.

What I Am Not Saying

I am not saying to never listen to others in the organization who may have an opinion about possible pitfalls. You need to listen to key people and weigh the evidence. However, consider the source of that information and whether the criticism is motivated by a sincere interest in improving things or, is their motivation to keep things the way they are by not tackling anything on by labeling it "risky?"

This motivation can be a challenge to determine. Naysayers do not announce they are naysayers. To properly weight a complaint, you must try to understand what might be motivating them to disagree.

Individuals, Not Groups

These types of decisions ultimately must be made by individuals not groups. Groups tend toward the average and lowest risk in all things. They work hard to minimize "perceived" risk in a manufacturing environment.

Your job as the project sponsor is to collect all the evidence in detail—all the pros and cons from all the various stakeholders and put it in a format that the management decision maker can ultimately digest.

Be Careful to Not Force Naysayers to Articulate Their Positions

I want to emphasize—asking people to articulate their disagreement in either verbal or written form will only cauterize their angst and give it more power. It will force the author to take ownership of the argument against your project or idea. Even if it is not that strongly held an opinion, it will become so after having to articulate it.

So, how do you keep from boxing people in a corner with negative perspectives? Be very careful about how and who you ask.

There is also a bandwagon effect going on here too: People like to pile on, and they love to be on the winning side of an argument. When they sense that the tide is negative or that "this idea will never work," many will also voice a negative viewpoint, even if they don't feel that way, just to have bragging rights that they "voted right" on this issue. "See I told you this was going nowhere."

Many naysayers gain power from voicing an unfavorable opinion. It feels good for some to take others down a notch. Not everyone is looking out for the best interests of the organization.

Or, for that matter, are they looking out for you or your projects?

Many organizations have become collecting points for such complainers. They are toxic to the organization and they create a culture of negativity, many light years removed from the positive early years of any business.

But, Build a Team That Will Win

We need to make sure the problem-solving team is made up of as many people favorable to the project as possible. You must include naysayers in the dialogue, but only if the main thrust of the project is relatively apparent and does not require deliberation—you want to save money on utilities. This is not rocket science; however, there will be many people within the organization that fight cost reductions tooth and nail for a variety of reasons.

A lot is riding on this effort. You must be careful about who you choose to lead the charge. One bad apple can spoil the whole bunch. And if we get a negative person onboard the team that is hell-bent on killing our project, it is likely they will be successful.

Plant managers often feel like they need to get buy-in from everybody. But once a vocal naysayer is in a group like this it is very tough to get them out of it. People then become afraid of not including the naysayer in all conversations for fear of not appearing to be democratic.

We all know of people who fit into this category and who common sense tells us should not be on project teams. Manufacturers and processors continuously include them anyway out of fear of appearing not to weigh everyone's opinion equally. I would suggest that there is nothing wrong with taking a stand and just flat out saying that a project is something we are going to do. I know times have changed. In today's modern consensus-based manufacturing and processing operations, many times people are afraid to call a spade a spade though. Managers, who are hogtied by shelves full of books by academics telling them to listen to everyone are finding it very difficult to make simple decisions. Things they know are right and just need to get done must be deliberated and voted on. Chances are, the agendas do not line up squarely with those of the managers or owners of the business.

All this is to say that we must have the confidence to move a little bit over the midpoint and push our agendas if we want to reduce costs. It is a helpful metaphor to look at our costs as if they were a leaky faucet. A leaky faucet gets fixed because it is visible. It reminds people constantly that it is broken and when things become an irritant like that, they get solved. However, with a project involving cost reduction, there is no visual reminder. The data are just too deeply buried in the utility bill's line items until somebody takes the time to shine a spotlight on them. I would say that is your job as the leader. You do not have to be the hands-on person to make sure the energy cost reduction project happens. However, you must choose your champion very carefully and require regular check-ins to make sure that steps are being taken and things are moving along. That sets the proper backdrop for getting things done.

The clearer you are as a manager or leader, the clearer your instructions will be interpreted by those who will execute them. In a way, this can be very liberating for the people on the shop floor. The fact that someone higher up has decided, and that decision is clear, offers a degree of peace for the people rolling out the solution. We all crave stability and certainty. The more we can give people that as leaders, the more successful we will be.

Does it Pay, Then, to Incentivize Your Champion Financially?

The simple answer is, it might. It depends on the person we are talking about. What are their personal motivations? Your goal is to keep things moving toward completion. This is something that takes some individual attention to figure out. Everyone has a slightly different metric for feeling successful.

Some people are highly incentivized at the thought of a bonus. Then, there are others who would consider it an insult. There is just no way to answer that question, in general, it is only answerable by understanding the person involved.

For the most part, people who work in a plant environment

are risk-averse. Because they are risk-averse, this is a filter that they run every decision through that affect you and your success.

CHAPTER SUMMARY

Plant insiders and utilities often kill worthwhile energy projects without due process. Plant insiders can be automatic naysayers, with knee-jerk adverse reactions to all ideas that are not their own. In the next chapters, we will discuss the tools and insights to combat this syndrome.

QUESTIONS TO THINK ABOUT

1. Establish a Baseline: What is the current total annual cost in each utility category for each meter? (chart the past 12 months electricity, gas, water and sewer)

2. Which rate is each utility and meter billed under currently?

3. What is the annual usage on each meter?

4. What are the rate options available with each utility?

5. Has the utility been contracted to discuss rate options and potential savings?

Chapter 2

Three Case Studies— Can You Really Save Millions?

The following three examples show money saved in several different industries. These discounts can be achieved by any business in most areas of the U.S. and Canada.

CASE 1: MULTIPLE LARGE PUMPS IN TEXAS —$180,000/MONTH SAVED

This company delivers vast amounts of natural gas and oil through their pipeline system in middle Texas.

The four utilities that service the clients' accounts were small and very easy to work with. They wanted to help their customer save money which is a huge plus. Each utility had a program that offered substantial savings in return for responding to power grid needs to reduce the amount of power used in late afternoons during the summer months.

Several of the utilities we worked with had contracts that had to be signed, but several offered the service as part of their regular rate. This was what was called a coincident peak reduction (CP) program.

In these programs, the utilities send out notices to customers alerting them as to when they think a peak in the utility's demand will occur. Through the CP program, the company was incentivized to reduce their operating load during those times so that they would be in lockstep with the utility's demand reductions.

This program helped the utility achieve its goal of lowering its overall system demand. They, in turn, were more than happy to pass along those savings to the company.

Let me say too that the company had known of this strategy long before I became involved. Their field personnel did not understand it. Thus, they were unwilling to implement the program because of an incorrect assessment of risk.

Specifically, what they did not understand was what the new level of risk *really* was. In their defense, this is not easy to grasp without a lot of work and attention to detail. Saving this much money requires a thorough understanding of the risk of any considered change as well as the inherent level of risk that the company is operating under currently. In fact, most companies do not realize that they are currently operating under a certain level of risk of downtime. And they also do not realize there is no guaranteed delivery of power in utility emergencies, as they usually think they are entitled to.

A significant challenge in this project was nailing down the coordination between utility communications and the field personnel who implemented those calls. Between the utility, their power provider, the customer, and the guys pulling the levers in the field, there were at least ten people who were in on the deal every time a CP power ramp back was needed.

CP curtailment notices were sent to the company and simultaneously to the people in the field. To achieve the goal of a CP program, a customer must go offline during the one 15-minute interval when the utility hits its peak.

However, you cannot know when the coincident peak actually occurred until months later when the meter is read and analyzed. So, you must pay close attention to utility notices as they try to anticipate that peak. You must react to all of them by reducing load during those periods. And, you must stay offline for a 15-minute period or 2 before and after it just to make sure that you covered the actual 15-minute interval when the peak occurred. Otherwise, you could respond to the utility-generated calls to interrupt, get back online too early and then miss the actual 15-minute peak for the month. In other words, this program must be done thoroughly and be done right or don't even try it.

CASE 2: PAPER MILL IN SOUTH CAROLINA
—$80,000/MONTH SAVED

This was a huge recycle paper mill that ran 24 hours a day, seven days a week, 365 days a year. It consumed massive amounts of electricity to drive material handling systems but an even more astounding amount of natural gas to supply heat for the process.

The company was in a rural part of South Carolina, quite a way off the beaten path. Most of the internal plant operations were robotically controlled, and the facility only had 75 people to run the operation that generated their million dollar a month power and natural gas cost.

On top of that, the plant was expanding. These factors had to be taken into consideration in crafting the best cost reduction plan of action.

After careful study of the gas and electricity infrastructure, it was found that the plant had the option of directly connecting to upstream suppliers without having to buy through the local gas utility. For the last mile of pipe, the local utility was adding a markup that almost doubled the unit price to the company for natural gas.

As an aside, local distribution companies (LDCs) are in most cases just middlemen for interstate pipelines. LDCs are given a monopoly over a geographical jurisdiction and charged with supplying power or gas to everyone in that area. The positive side of this proposition is that everyone gets served. The negative side of the proposition is someone must pay for it. Often, the cost of supplying power or gas in remote locations is paid for by local businesses and industries in the form of higher rates to them as we see in this case. Thus, businesses often carry the rate burden of the utility supplying residential users power or gas—regardless of whether it is economically feasible to do so.

That was indeed the situation with this LDC—natural gas was being transported in through hundreds of miles of interstate pipelines and then moved only a few thousand feet by the LDC to the recycling facility. For that short run of pipe, the LDC then charged the company over $2/dekatherm, which was six times the price of the long-distance interstate pipeline transportation.

The option we created for the company came through the mechanism of FERC Order 636. Order 636 allows individual companies to connect directly with upstream sources or interstate infrastructure.

"Open Access" has allowed private and public companies to cut out the middleman and buy direct, taking control of the transportation portion of their gas cost. Due to rights of way rules, it can be a lot of effort to size up such a project and install a line tap and pipeline, but for that effort, a company can cut their utility bill by a drastic amount.

Being willing to get into the details and size up this scenario was essential to success. The plan was, finally, to construct a pipeline from the interstate gas pipeline directly to the plant and cut out the LDC's participation in the billing altogether.

This project involved some private property owners understanding rights of way because the pipeline would have to cross some streets. This level of detail was time-consuming but not difficult to do. Pipelines and cable companies regularly work within these same rights of way. That is because FERC Order 636 forced state and local governments to open access to utility trenches to anyone with the means and need to install their distribution facilities—all in the name of freeing up interstate commerce.

Once the project was developed, we informed the local distribution company of what we were doing and the date the company would no longer need their expensive services. As you may well imagine, the LDC was not pleased. They evaluated the project and measured the strength of our intent. An offer was made to continue service at a more reasonable price, and all left happy.

CASE 3: SMALL OPERATION THAT SAVED 40%

This company was located on a small plot of land in the middle of a large city. Assembly operations took place in over

40 buildings. Each building had its own electricity meter, so the company was receiving 40 bills a month from this small utility, each coming into accounts payable at a different time.

Even worse, each meter was generating a utility invoice for a different unit cost for power. One meter would average $.11/kWh and the one right next to it would be $.06/kwh. All within a stone's throw of each other. Doesn't sound too fair, right? The buildings were only 40-50 feet apart and all on the same small piece of property (2 acres). Could the energy be worth more in one building than it is in one right next door? I don't think so.

But, meters are the way utilities expand their empire—put a new meter on every new location and every building. In fairness to the utility, most customers say they want a separate meter on every process because it requires the least out of pocket money for them to get started. And, having a separate meter makes the operation easier to track. Putting the utility meters on everything allows them to collect a "customer service charge" for each separate location.

The reason the utility's motivation for installing separate meters is essential to know is that it drives costs up on a lot of electric rates. Rates are often structured so that the most expensive $/kWhs occur at the beginning of the month. Following that, the rates are stratified so that energy costs less later in the month. What that means for our purposes is that if you have a lot of kWhs from running 40 different buildings as our example customer had, the typical utility model requires you to run that usage through the most expensive hours on the rate repeatedly at each location.

To simulate the cost and benefit of centralized metering, we first compiled all their metered data for a month and simulated running that data through the same rate once, instead of 40 times with the 40 different bills. The difference in cost was astounding. The savings shown in the analysis was over 40%. The utility was presented that information, and we worked together with them to meter the entire facility at one point so they would then invoice the customer only once a month.

Creating a single meter point at primary voltage was a straightforward solution that yielded an enormous and fast cost reduction for this client.

Consolidating metering can be a robust solution for any facility with multiple meters like this. The 40% reduction was an exception because of the number of meters, but you should typically expect savings on the order of 10-20% from consolidating or "totalizing" the meters.

"Totalizing" the meters is the über-concept, and it means leaving multiple meters in place and then reading the metered data simultaneously and combining them electronically. It is usually not unduly expensive, only requiring the purchase or rental of devices to communicate and accumulate usage.

Always consider both ends of the spectrum—combining and disassociating the meters though. Either strategy can yield positive cost savings results for a manufacturer.

There can be advantages to combining and to splitting up the electric load. Disassociating can often be a compelling alternative if usage conditions have changed. If production lines were shut down or equipment is taken out, splitting things up can be an especially good strategy. It is simple for your utility to help you with an analysis of the options available for splitting up metering.

THE BAD THING ABOUT CASE STUDIES

The results that are achieved for any one company are highly contingent on the details to that company's operation, the industry they are in, the utility they are tethered to, and their geographic area. There is no such thing as a cookie cutter solution in energy cost savings. The answers here are as individual as snowflakes, and that is why I am always hesitant as a consultant to give people case studies.

People usually think they can either copy the case study exactly or rule it out completely. Neither is the situation, however. There will be elements of case studies you read that will apply to you and your operation and some that do not apply at all. The bottom line is that you must take them all with a grain of salt. I believe you should view the case studies as a benchmark. There are always better ways to approach things, and in my experience, there are still more ways to save additional money if you are persistent in creatively initiating new potential billing scenarios for the utility to consider to vet.

An example of one of these inequities that can change the potential of a project is if you happen to be a service provider and not a manufacturer, the options you have at your disposal change dramatically. Utilities do not market as vigorously to certain classes of customers as they do to others. It so happens that most utilities market aggressively to manufacturers. Why do they do that? Because manufacturers consume power consistently around the clock. This round-the-clock work schedule syncs up nicely with the internal economies of scale of a power company.

If you operate your facility only during daylight hours and not round the clock, your options will be very different. Utilities will view you and your business as a "partner" the more closely your facility runs around the clock. If you require the utility's "product" all day and all night, you will garner the most attention from their reps and the most options for cost reduction. You will find utilities bringing out the best of the best to cater to you if you have the potential to consume power this consistently.

If you are expanding your facility, your status will also quickly change to that of super-customer. Because utilities make more money by increasing their asset base, they love conversations about your potential expansions and growth. Even the mere suggestion that you might develop your manufacturing capabilities will open a range of options that you currently do not have. I am in no way saying you should enter into such discussions with your utility if you do not plan to expand, but if you do have even conceptual plans in the works, you need to let all your utilities know at an early stage. This will get them to work coming up with ideas to incentivize you to expand and grow. Their incentives will make your decision to expand all the easier.

Utilities are also as different as night and day in the way they pursue new business. Even ones that are right next to each other can be radically different since they are not forced to compete with one another. "Investor-owned" utilities (IOUs) have a different set of rules that they must operate under than Electric Membership Cooperatives for example. IOUs tend to be very sophisticated when it comes to working with state and local governments to offer manufacturers big packages of lower prices and tax benefits in exchange for expanded facilities. EMCs can be aggressive as well, depending on the state, but they are not as used to being competitive as IOUs, so the process of getting a lower bottom line cost may not run as smoothly.

Finally, your geographic area can offer you additional possibilities to help write an even a better case study for you. Not only at a local level (county and city) but also at the state level. Every state has a different attitude when it comes to offering incentives and attracting various customer classes. Some can be very aggressive in wanting your business. Those states and local governments see an expanded manufacturing base as bringing in additional tax money and other benefits through hiring additional employees for the expansion. Some other states are not so far-sighted and do not offer incentives at all.

What does this discussion of case studies and their real meaning mean for you? How should you interpret the case studies that I am showing you and others you may see? I recommend that you look at them all through the prism of skepticism I have just described. To maximize your plant's own "case study, " you must look under all the rocks for additional savings. You can't copy someone else's case study and do that.

Case studies should point you in the right direction and should motivate you to act, but they never give you an exact roadmap to success. There are just too many other variables in your company's cost savings formula. Your options could be better than those described in the case study if you are creative and relentless in pursuing them.

WHAT HAVE WE LEARNED FROM THE CASE STUDIES?

So, what is the commonality among our three examples? None of the solutions was evident in the beginning.

Your utility rep most likely will not be up to speed on many of these strategies. Even the upper ranks of your utility's management may not be familiar with them. You might not even be able to find out a single bit of information on these strategies from a Google search of the topic.

The main thing I want you to grasp here is to not take utility billing at face value. Apply common-sense reasoning to the utility cost reduction process and do not ever think, "if this were a great idea, someone would have done it already."

Because of their history as government entities, we too often accept what utilities say as if it is the order of law. We don't push back much when they tell us an idea won't work or is impossible. They often merely ask us to read the rules they have attached in the email. They will tell you, "You are paying the best rate!"

The best relationship you can have with your utility, though, is a symbiotic one. Whatever benefits your business must also benefit your utility in some way, or it is not going to happen.

In no way do any of your interactions with your utility need to be hostile or confrontational. That will get you nowhere. Working with them is not like any other vendor or supplier relationship you have. That's because they see you as being a captive audience. While they would prefer you to like them, they know deep down you don't have a choice from whom you buy.

What does that mean for you and the way you go after achieving savings? It means simply you must realize that you must operate within the ecosystem of the utility's rules and regulations to get things done. In most cases, it does not benefit you to "leverage" the utility/customer relationship—that is, to threaten them that you will leave if you don't get your way, etc. That will only sour any good vibes that could work in your favor.

Utility employees are usually very loyal to their organizations, and any evidence of disloyalty will be taken as a threat. You will get nowhere with them if that is the interpersonal dynamic you project.

Utilities do like to fix problems for companies that ask politely. It pays to be a square peg not fitting into their round hole. Just be a polite square peg.

So how do we sum up the proper attitude? Approach your utility with the attitude that you want to understand their system. You can find ways to use what you learn to further your cost reduction goals.

Entering a dialogue about the way their system operates, with the objective of reducing your bill is a subtle approach. Realize there is no free lunch.

It pays to consider your current position carefully. In almost all cases I find the end-use customer is currently exposed to more liability than they realize. And they are not getting a benefit or payment for it. That is where a cautious and honest risk assessment and a sober rebalancing of liability is needed. It could be that you are paying extra for a lot of redundancy and "insurance" in the form of averaged pricing.

I want to reemphasize that you need to be on a mission with your utility to use their system to your benefit. While it is unlikely the utility or your rep will be able to answer all your questions, making them look good in their boss's eyes is always the right thing to do. There is no seldom an upside for you in going around the rep to get things done faster. With a bureaucratic utility, it is essential that a rep become a source of information for you. By including them in the dialogue and showing him or her respect you are setting the stage for them to help you reduce costs.

If the utility feels you are just going through the motions and not as much interested in cost reduction results as you are in checking off the block that you at least tried, they will gladly accommodate with a no. But if they sense you really need cost reductions to help keep your business viable, they can mobilize resources internally to make that happen. The utility will enjoy being a visible part of your success.

Be open to new options for saving money that you haven't

thought of. The utility may bring those up. It does not have to be your idea or the idea that you started with. When a utility feels they can save the day and keep you, their customer, happy, (and keep jobs in the community), by getting on your side of the table and getting creative, they can become very creative.

Some people who work at utilities are very solution-oriented. Often, we start out thinking the solution to the cost reduction problem lies in one area with one solution or idea, only to find out through the process of interacting with the utility that many other ways to save have unfolded. It is a good strategy, therefore, to "stir the pot" with one specific idea to get the process started.

For instance, with my case example, the original idea was to build a natural gas pipeline to connect the plant to the upstream interstate pipeline. By the time the project finished, a compromise was worked out with the utility to make the full implementation of the pipeline unnecessary. We still achieved almost all the cost reduction goals we set out to achieve. You must stay open to other ideas in this process if you want to save a lot. Your utility may well say, "we can't do that but we can do this" …which is fine because you are going after paying fewer dollars, not specific solutions. Be agnostic to the source of the savings.

CHAPTER SUMMARY

These three case studies from diverse sources have detailed over $2 million in annual savings. These are not isolated cases, and you can achieve the same level of success by following the approaches we are discussing. The best results come from not having a death-grip on your original idea and being open-minded about where the savings come from.

QUESTIONS TO THINK ABOUT

1. Have I checked with industry associations to find case studies available for my industry with each utility type?

2.   Have I contacted the person who authored the case study to learn more?

3.   How do the case studies I have read apply to my operation?

4.   Have I contacted my utility for case studies in cost reduction they have (white paper and verbal)?

5.   Is anything stopping me from taking action?

Part II

Tools for Demand and Cost Reduction

Chapter 3

Demand Response/ Interruptible Service

Demand response/interruptible service is a category of cost savings programs that will pay you handsomely and in multiple ways for being "first in line" for a possible emergency power curtailment. Leaders that understand its power feel like they own a personal moneymaking machine.

But, many have heard about this type of program and have gotten just enough information about it at the *50,000-foot or concept level* to dismiss it summarily. Not all utilities and power grids have interruptible programs, but if yours does, we'll look at why you should first thoroughly understand what it means locally, and then find a way to make it work for you.

At a high level, interruptible service or "I.S." often gets breezed by or even completely ignored thanks to the deep research necessary to evaluate it properly. There is no one-size-fits all plan. I.S. must be understood deeply for it to make sense. If you go to the trouble, it can be one of the most formidable tools in your arsenal to cut costs. That's because I.S. has everything a commercial or industrial energy user wants in an energy cost reduction program:

- Big financial rewards for you ($25,000-$120,000 per 1,000kW)

- Reasonable contract term horizon (usually, less than 5 years)

- Participation in such a program cuts carbon emissions

THE ORIGINS OF INTERRUPTIBILITY

Many years before the first forays into deregulation that began in 2000, utilities had an ad hoc system of taking care of their power supply dilemmas. When they experienced a power emergency, the plan of action was to call a short list of large manufacturers and ask them to reduce their loads to shut down the crisis. Some would and some wouldn't reduce their loads based on their production schedules at the time. The utilities, of course, always had in their back pocket the ability to involuntarily curtail their customers' loads if it was an actual emergency and maintaining the integrity of the grid hung in the balance.

In the 1990s, utilities formalized these loose agreements with interruptible rates. These rates paid the customer for complying with the requested curtailment, making it a lot more likely they would do what they were asked to do in emergency situations.

The deal was industrials would stand ready, but what industrials were really expecting was for the utility to supply continuous power with zero calls to interrupt AND still pay the discount offered. In other words, the once-discounted program had become the expectation of a discount and no longer had the power to influence behavior as it was first designed.

In 2000, California was the first state to deregulate. The system of pricing the electricity commodity was long conceived but extremely flawed. Prices soared. The expectation was that when prices soared, manufacturers would fall offline in response. The laws of supply and demand would kick in and grid pressures would be relieved. That didn't happen, however. Many just kept running their operations, and prices crept higher.

The non-compliance penalties that were part of power supply contracts at the time were of course invoked. Many manufacturers simply refused to pay them. Utilities sued and it became a big mess. California has since re-regulated, and Californians suffer under some of the highest power supply prices in the country now (over $.30/kWh).

This was the perfect stage for the advent of demand response, a milder derivation of the interruptible rate, to enter. Demand response programs addressed the problem of compliance by supplying the market with a pooled power. That means that there are many participants in each pool of power that can be dispatched or called on to curtail in times of grid emergency. The reason a pooled system works better for the end-user customer than the old system of every man for himself is because in general not everyone in the power pool must perform for the system to work and relieve the load on the power grid.

DETAILS YOU NEED TO KNOW

I understand those who throw up their hands and give up on this topic. Interruptible service rate sheets and formulas are usually eye-crossing in their complexity. They confuse people and make them shake their heads with legalese. To add to the confusion, there is a wide discrepancy between what these rate sheets state and the way utilities behave in real emergency scenarios and in real penalty adjudications. To top it off, there is nowhere really to turn—utility reps seldom understand them either.

Simply put, interruptible power rate sheets universally make the programs look like sharks, but most of the programs are like guppies. They are only made to sound way worse than they are to scare you off. Utility "Rate Departments" spend years drafting them. They are written with loads of boilerplate about stiff penalties for non-compliance and such to make you feel like the Armageddon scenario of hundreds of hours of interruptions they paint for you will come true.

Here is one little tidbit of information that utilities typically leave out when they explain these rates that is critical to understanding how beneficial they can be for you and where you stand now.

In most cases, your utility can interrupt you right now, without warning, if they feel it is required to maintain the integrity of their pow-

er grid. In such an emergency, your utility would try to give you notice, but they sure don't have to. The key to understanding the strength of interruptible service for your business is to know *how close to the cliff you are already standing* and how much money is in it for you for being willing to stand a few millimeters closer to the edge of that cliff.

It is very likely that your utility will not have had any actual interruptions at any time in the recent past. If there are no interruptions called by the utility, they may ask you to perform a "demonstration," or a simulated interruption for 1-2 hours but that may be all you have to do to earn capacity credits.

The first reason utilities say they need a demonstration of your ability to interrupt is they want to know for sure that you can drop your load when and in the amount you promised. Secondly, there is increasing pressure from customer groups within states that see interruptible programs as de facto discount programs that utilities give their big customers. These groups see the I.S. programs as having no real value to the broader group of utility customers in this era of low-likelihood power reliability issues. These groups feel the utilities are just giving away the money and getting nothing for it. That is why a demo request is a strong signal that it is unlikely you will experience actual interruptions.

WHY UTILITIES HAVE INTERRUPTIBLE SERVICE

I know you are thinking, the term itself somehow sounds troublingly oxymoronic—"interruptible *service*." Shutting your power off sure doesn't sound much like service.

Despite the frightening sound of the phrase, interruptible service power has a virtuous vector. It was created as an attempt to save the environment.

Back in the late 1970s, the U.S. economy was expanding rapidly. Utilities kept up with the expansion by building a lot of new power plants to stay one step ahead of the growth.

Utilities, being profit-driven businesses, did not mind the hectic economic expansion one bit. That was the way utilities had traditionally grown—by adding assets or power generation equipment in their service territories.

This is the way the utilities' business model has always worked: Every time their consuming public hit a new peak demand level in the summer, the utility petitioned its local Public Service Commissions (PUC) to build another power plant to meet that demand.

But, nevertheless, utilities would build new power plants and then turn right around to state public service commissions and ask for approval to recover their investment and then some from the consuming public, regarding higher rates to all consumers. The utilities were already guaranteed a return on their investment by law. Not a model for how to ensure efficiency or low costs but a profitable and repeatable model for the utility.

After years of fattening utilities' wallets at the expense of the consuming public, state PUCs had finally had enough and told them, *"Instead of continuing to build new power plants every time a new summer peak is set, why don't you go back to your largest customers and ask them to reduce their power occasionally to compensate for the occasional peaks?"*

Thus, interruptible service power was born, and the plan began to proliferate everywhere new generation capacity was proposed in both the U.S. and Canada.

For the most part, we are a trusting species. For most of us, the default position is low stress relationships and not rocking the boat. While you don't need to be confrontational to be effective, challenging the herd mentality is necessary to evaluate true risk.

Adding the risk of interruptible service to a business already stressed from seemingly uncontrollable factors seems wrong to many managers. This is a rationalization to avoid the stress of challenging the prevailing mindset. These managers give up quickly on I.S. at the concept level. I can understand that but you cannot afford to allow it to define the way you do business inside a manufacturing plant. For example, if you read the utility's rates without more local information, you would likely conclude that adding interruptible service is a bad idea.

It is ironic that most of the manufacturers that can benefit the most are the most afraid of taking action. The idea of interruptible service is always scary without more detail and raw data to shine light on the actual level of current risk and the alternatives.

It's not your fault, though. The frightening nature of I.S. is partly a contrived product manufactured by the utilities. Lawyers experienced in rate design write their interruptible rates and they purposely make them sound litigious and draconian, hoping to dissuade companies from signing up. If you read through a power rate and conclude that the utility's intent it is to explain and clarify, please call me—I have the Brooklyn Bridge for sale.

Think of it from the utility's standpoint. Why would a utility push a product on you that is guaranteed to decrease their revenue? It is a little like a manufacturing business telling customers, *"Hey, Mr. Customer, would you like to pay less? We can offer you a discount if you simply agree that we can stop production and not send the electricity to you whenever we feel like it."*

On top of that, we humans always say NO when something is not clear to us. Utility rate writers are no slouches when it comes to understanding what emotions drive people's actions. They take advantage of our fears and weaknesses as they weave a high-level understanding of human psychology into their rate documents. They do not really care if you understand them and, thus, they make them eye crossingly vague and fear-inducing.

Interruptible power cannot be adequately or accurately analyzed without a lot more work on your part. How you will do this is much like the way a life insurance company sizes up you as a risk. The insurance company must deeply understand the likelihood that you as an individual will die within a certain time period.

How old are you? Do you smoke? Are you college educated? You know the questions, and the list goes on and on. The insurance company asks enough questions about you to understand the gap between your desire for life insurance and your real NEED for life insurance. The insurance company, of course, only wants to insure you if you don't need it.

Why Does Your Utility
Think They Need to Interrupt You?

Generation plant emergencies cause most interruptions and the genesis of most power emergencies is extreme weather. Very high and very low temperatures force utilities into overdrive because their generators must run at full tilt to supply homes and schools heat and air conditioning. Here's the scenario: The maintenance normally preformed to keep generators running can't be done because the units are running wide open all the time. The generators break and they trip offline. When the generators trip offline or are taken offline to keep them running through required maintenance, the utility usually first tries to purchase supplemental power from neighboring utilities and if that can't be done, they look to their other options for supply power to make it up, like interrupting your operation.

How to Analyze Your Options

There is a lot you must understand to analyze interruptible power locally. Just like the insurance company analogy, you must ask a lot of questions if you want to assess the real risk the interruptible program presents to your plant.

All I.S. rates and demand response programs are created equal. Depending on the utility, interruptions can happen anywhere from never to 5 times a month. There are significant differences between them, and most of them depend on the region you are in.

If you do not have backup generators, you may not want to opt in for a program that requires you to shut down frequently.

Here is a checklist of starter questions that *must* be asked of the utility, PUC, and others to understand if this might be an option for you:

• How many interruptions have been called over the last ten years?

- When were those interruptions called? (Date, time, duration.)

- What were the circumstances which caused the need to call those interruptible events?

- Is there generation available now at your plant? How much? Is it enough generation to supply power to the whole facility?

- How much time during a given month or week is preventive maintenance done on production equipment? (This is a window of opportunity for manufacturers wanting to take on interruptible service because plant maintenance can move into preventive maintenance mode during any power interruption.)

- Is the utility making or planning changes to their power plants or delivery infrastructure that would make an interruption likely to occur?

ALSO, YOU NEED TO MEET THE I.S. PROGRAM MANAGER

It is essential that you try to talk with the people involved in the day-to-day running of the I.S. programs you are involved in. In the conversation, ask direct questions about what internal metrics would trigger interruption events and what hoops need to be jumped through for the program manager to "dispatch" an event.

Not just interruptible power, but all systems boil down to the actual people and stressors involved in executing them. Understanding the human side of the individual interruptible rate and the trouble the utility must go through to call an interruption will help clarify the likelihood of one happening.

Also, since you have met the person and told them a little about your operation, it makes the impact of shutting off your power a little more real for them. Now, he or she is not just interrupting the power at some nameless factory, but he or she is shutting off the power of someone they know whose employer is an asset to the community. This alone could be a strong interruption inhibitor and may keep the utility from calling an interruption. As manufacturers, we would prefer the utility to look for other options to solve its power delivery problems.

AN EXAMPLE OF INTERRUPTIBLE SERVICE THAT WORKS

Oklahoma Manufacturer

One of my clients had locations in Oklahoma and Alabama. Both of their utilities had interruptible service programs available. A detailed analysis showed us that neither of the utilities had called for an interruption in the past two years and, in the case of the Alabama utility, there hadn't been an interruption event called in ten years.

The company signed up for both programs. The Oklahoma utility program pays the company roughly $450,000 per year and the Alabama program, approximately $180,000 per year.

In the two years they have been on these programs, there has only been one interruption collectively. The interruption was caused by a transmission level power line that fell in an ice storm. In other words, an interruption would have happened anyway, and my client would not have been compensated in any way.

This is lack of need to actually interrupt, which is common with these programs. Power pools or grid networks are not stressed now. In many areas of North America, that is thanks to the advent of green energy supplies, such as wind and solar energy. These additional sources of power generation have reduced the demand on the grid and have subsequently made reliability interruptions much less likely.

ECONOMIC CURTAILMENTS

Since many demand response programs are not actually needed to relieve pressure on the grid as much as they were at their inception, some utilities have found a way to make the most of the rare structure that is in place. Utilities have found they can sell power to neighboring utilities at a premium during peak times when power costs are at a premium. Since the power must come from somewhere, the utility will ask interruptible program participants to curtail load to sell the power they purchased ahead for you to someone else at a higher price.

This is not happening everywhere yet, but utilities are becoming more comfortable with this type of arrangement.

The Attitude You Must Have in This Analysis

To lower your costs, you've got to think differently. How *can* I use this program? Not, how can I find the weaknesses and shut this down so I won't be bothered by it anymore. You must be willing to examine the relative risk fully and objectively.

With stakes often in the millions of dollars, this is the only energy cost-cutting mindset that you can afford to take. You must be agnostic to the hassles. Objectively look at how you can work around them.

If you take this seriously, I have given you a lot to do. It may take months to get the proper level of understanding you need to evaluate interruptible service the right way. The only way to approach a problem of this magnitude is to break it down into bite-sized chunks.

It will be difficult to stand behind a program that you are only marginally familiar with. That is why I advocate drilling down to a great level of detail over time. A cursory analysis simply won't give you the confidence you need to pursue these programs in the face of risk-averse naysayers.

What Is the Next Step?

As we've said, while the I.S. programs you seek are not hidden, they will not be seeking *you* out. That means you have some work ahead of you. But the payoffs can be very large.

Do not expect the utility to jump in and take on your level of interest in this topic. Expect them to throw out dire warnings of cataclysmic disasters if you are foolhardy enough to make your plant exposed to interruptible power. Remember, the utility people you will talk to are very conservative by nature and usually always opt for the lowest risk option.

Understand your current *real* level of risk thoroughly. In most cases, utility interruptible programs are the last step before the utility calls for a rolling brownout. A rolling brownout means the utility can and may pull the breaker at your plant and shut you down without providing you compensation.

In the final analysis, by signing up for interruptible service you may not be exposing yourself to very much additional risk at all. That is what we want to discover, though, through your in-depth homework. The incremental risk you take on could mean millions to your plant over time.

What Are the Best Reasons to Do Demand Response and Interruptible Service?

1.   Demand Response will put a significant amount of money in your pocket. Besides getting that big deal from heaven (or from hell) with Walmart, there is not a single thing you can do as a manufacturer that will put more money in your pocket any faster for doing less work.

2.   Demand Response does not cost any money. Most energy cost savings cost money. If you want to save in lighting, for example, you will have to write a big check. Demand Response has zero set up cost. All it takes is your time to analyze the situation properly.

3.   Demand Response benefits your fellow humans. No one wants to live around the corner from a new power plant. These programs reduce the amount of power needed to be supplied at peak times and thereby eliminate or delay the need for utilities to build more of them.

4.   Taking calculated risks has a positive effect on other pro-
      cesses in the organization too. This will motivate people to
      analyze and take on other reasonable risks in other places
      in the company. Nothing much ever happens in life unless a
      shift occurs in the risk-reward balance.

CHAPTER SUMMARY

Demand response programs can put a lot of money in your
pocket for minimal effort You must be creative and figure out
how to make these programs work.

QUESTIONS TO THINK ABOUT

1.   Does my state allow demand response programs, and does
      my electric utility offer an interruptible rate?

2.   Have I located 3 alternate providers to contact for pricing?

3.   Is there backup power generation at my facility?

4.   How many likely events per year will there likely be and
      when will they likely occur?

5.   Do I have the ability to shift scheduling perform required
      machine maintenance during demand response events?

Chapter 4

Thoroughly Understand Your Relative Risk to Win with Interruptible Opportunities

For our purposes, it is not enough to do an everyday surface-level analysis of risk. You must dig deeper and understand not only a detailed picture of the risk and reward a new energy program offers you, but also what your current level of risk is.

This is what I mean by "relative risk"—how much risk are you exposed to now and how much risk do any of your new energy cost reduction options present you?

As a backdrop, let's discuss the nature of the power markets and the kind of risk their operation presents to a manufacturing business.

IN DEREGULATED POWER MARKETS, THINGS WORK LIKE THIS...

In most deregulated power markets, there are a small number of actual suppliers of energy and a whole lot of brokers or resellers.

Back in the late 1990s, the country's power delivery problems were blamed on the lack of free enterprise in public utilities. Proponents of deregulation argued that the reason power costs were so high and the system was so constrained was big utilities had cornered the market and had become lazy. They were making their profits and the public service commissions around the country guaranteed those profits.

Some utilities figured out how they thought they could make money at the deregulation game and many got on the deregulation bandwagon. It's hard to argue against a free market.

California was the first, and to give you a feel for how deregulation has gone, has since RE-regulated 17 other states deregulated from 1999-2003. All the other states that had plans to deregulate have shelved them.

The reason is, deregulation only shifted costs from one class of customer to another. Prices for power went up in most cases, not down. And, on the consumer side of things, the system took one of the smaller expense items and introduced a substantial new hassle.

This is how it works in a deregulated power market—your power purchases are broken down into "generation" and "delivery." The generation services of public utilities were spun off in those 17 states that embraced deregulation and were sold to private companies who run them and offer the power commodity to consumers and businesses.

So, What Options Are Available to Reduce Costs?

The fact that utilities market to customers that have round the clock requirements is good news for all other consumers. Utilities entice high load factor customers to locate in their service area to bolster power plant utilization.

If you also fit a high load factor load profile as a manufacturer, you can benefit from that type of rate, too, even though the utility didn't create the rate with you specifically in mind. What should you do first if you want to reduce generation or supply cost when you are operating in a deregulated state?

The first thing to do is get copies of your last twelve months of invoices for each meter. Understanding the big picture of power consumption is always the best place to start. Then, find out which rate each account is on and build a spreadsheet showing usage, demand, and any other item that the utility is charging you for that may be controllable.

That is easier said than done, though. Many times, utilities, especially smaller ones, either don't list critical information on their invoices, or they list them in "utility speak" and make it difficult to understand. Calling up and asking the utility for help over the phone probably won't provide a lot of support either. Most of the people you talk with on the phone do not know very much.

If the rate does happen to be on your bill, it will usually be in a nickname of sorts—PG&E, for example, calls one of their rates "E-32 TOU." However cryptic they make it, once armed with this information you can go to the Internet and look for an alternate list of rates for your utility. Utilities usually segment their rates into residential and business, so you will not have to sort through all their rates if you use "business" or "commercial" in your Google search.

To cut costs, we will be looking at several factors—first on the list will be your load factor (LF), which is a good indicator of how often and for how long you run your operation. A quick and dirty calculation method for estimating the impact of load factor is a ratio known as (HUD), or Hours Use Demand. The way to calculate HUD is to divide the total kWh/kW. The higher that number, the better deal you will usually be able to get from your utility.

A load factor or HUD score of 600 or higher will be a home run for you because it indicates your facility runs pretty much around the clock. Electric utilities always market great new rates to the high load factor class of customer because their run schedule helps them keep their power plants selling electricity all the time.

Having a HUD in the 300-400 range is also very good, though and means the facility will also likely qualify for cheaper "time of use" rates for many utilities, but it all depends. There is no set formula.

Even though they are not the top of the food chain of rates, time of use rates can still be a powerful cost saver. They were developed in the 1990s to entice customers to consume power in cheaper, off-peak periods. It just so happens that many facilities benefit from TOU rates without shifting any load to off-peak at all, because their consumption patterns already fit better into that rate schedule (even though your utility may tell you otherwise).

How Should You Approach the Utility about Potentially Lower Prices?

The best way to contact your utilities is via letter or email, not over the phone. When you make the conversation in writing, you can articulate precisely what you want your utilities to do and it can't be confused as things often can when on a phone call. Also, asking them in writing allows them to forward it on to the right person.

Furthermore, you don't know who you will get on the phone if you call a utility. (Most likely, you will get stuck at a lower level if you insist on calling them.) In many cases, the person who answers the phone usually just started work there and knows very little about the rates and things you can do to cut your costs (although, they will say they know). That silence after you ask them a question is them shuffling papers to read you an answer off a prepared Q&A sheet. This is not what you need when you are dealing with high-dollar issues. The savings often lies in the details, and a generic explanation simply will not do.

The letter or email you write should ask for optional pricing programs available and should name the programs—be as specific as possible. As we've discussed, the available optional rates that you can include in the letter are on the utility's website nine times out of ten.

To Whom at the Utility Should I Send the Letter?

I recommend sending the letter to a C-level executive if you do not have an assigned representative from the utility. Why so high up on the chain of command? They can pass it on to whomever they need to get the job done. You will get higher-quality answers and a more "motivated" response when a C-level personage is involved.

For an industrial facility, the utility usually will have a specific representative assigned to the account if the annual spend amount is large enough and then you can just approach him or her directly.

How Much Can I Expect to Save?

We find that most manufacturing companies will save on the order of 10-15%—that is, if they will benefit from an alternate rate or program.

Can the Utility Put Me on a More Expensive Rate Just for the Asking?

No way. This is many people's fear, and it is entirely unwarranted. This fear keeps many from even trying to lower their costs. The truth is, utilities just don't have the time, resources, or desire to punish their customers. By and large, your utility wants you to survive to pay your bill another day. They will work hard to help you if you supply them with a good reason (expanding operations, goodwill for the utility, you are kind to them, etc.).

Utilities and their representatives won't do anything that will make their next visit to your plant awkward or painful. With frequent outages in some locations and other power quality problems, it is no surprise that utility reps are on the defensive even. Your request to look for alternate ways to save within their system will not be met with hostility. You are giving your utility a chance to help you—regardless of whether they are able to lower your costs or not.

The C-level or utility rep you deal with wants all the pegs to fit in the right holes for you as their customer. He or she wants you to rate them and their utility highly when the next "independent survey" comes lands in your inbox. Your positive comments often insure their raise.

Utilities in the past used to think that you, as the customer, owed them something because they were gracious enough to supply you power or gas. That attitude seldom exists anymore.

THREE KEYS TO UNDERSTANDING BILLS

The first thing we need to do is understand the way utilities build their rates. The language utilities use to explain their rates is stilted and purposely hard to understand. That is, lawyers help write rates.

You must be willing to dig deep down into the details to understand. That means asking a lot of questions. The details are where the gold is buried so you can't let a question be answered in a way that doesn't clarify things. You are entitled to understand, and you are entitled to answers to your questions, so continue to ask until you are clear on what is going on and all the line items on your bill.

The root answers you need are often not found at the utility, however. Often, you must go on a personal fact-finding quest far beyond utility rep-level understanding if you want to save big dollars. That can mean going upstream of the utility and their representative to the grid operator.

The basic structure of utility billing, however, is not all that complicated and consists of three main components; how much power you used, what was the maximum you used at any one point in the month, and the amount you must pay.

The main things you need to focus on at first are:

- Demand—how much is it and how is it calculated?

- Consumption—how many kilowatthours?

- Rate—what is a rate called and can I change it?

- $/unit—how much am I paying relative to other accounts I have?

Once you have those things noted, you can begin to formulate questions. The key to savings is very often lodged in the demand cost. That can be half your bill or even more. It consists of a lot of different variables that you may be able to control inside your facility. For example, if you turn on all your equipment at one time, that can spike your demand and drive up your monthly cost. One 15-minute period of lack of demand oversight can increase your monthly power cost by 50% for the entire next year.

The customer was billed for that amount of load even though 99% of their demands were much lower over the rest of the month. Demand spiking is something that you can control. You need to be hyper-conscious of your demand always. We will talk about how to do that later when we discuss energy management options.

The demand cost reflects the maximum amount of energy you used that month or in the last eleven months (or, your contract amount). The way the utility looks at it, what they expect you to use is the maximum amount they could be called on to supply to you during the month. That is a cost, however, over which you have a lot of control without spending money.

The demand cost can also be stratified between on and off-peak. It can also be stratified by the utility's demand and their generation supplier's demand. Understanding thoroughly the different ways your utility charges you for demand is a significant key to lowering your costs.

WHY WOULD A UTILITY CARE *WHEN* YOU USE YOUR ENERGY?

Utility math isn't straightforward. It is important to know, though, that most utilities have a fire sale on nights and weekends. The reason for that is because generation economies of scale require them to run around the clock. However, not all businesses consume energy at those times naturally. So, that opens a great opportunity to save for those businesses who will consider changing their operating hours to accommodate their utility's pricing rates.

You can save a bundle of money on your power costs by adapting to utility off-peak pricing schedules. By modifying your operation to consume most your power on nights and weekends when power demand is otherwise low, you can cut your bill in half. Even if you only were to shift a portion of your energy consumption to these "off-peak" hours, you will save money.

A lot of manufacturers I have worked with feel like they could never get their crews to go along with the shift in hours needed to benefit from a time of use pricing products but I say, nothing ventured equals nothing gained. It may just be that enough of your workers are ok or even prefer such a change.

UNDERSTANDING HOW YOUR UTILITY BUYS POWER

Critical to understanding your power costs is understanding how your utility generates or buys the electricity they send to you. They will welcome the chance to tell you more about this. Your rep may not understand the big picture, but will be able to connect you with those in their system who can explain these things. I know it is time-consuming, and you have important things to do, but there will be a huge payoff waiting in potential cost savings if you take the time to understand how their system works. The key to scoring big and saving a lot of money is first understanding what makes your utility tic and learning how to work inside their ecosystem.

Once You Understand the System and How Everyone in the Loop is Incentivized, the Rest is Much Easier

Learning the details of the way your utility gets its power will tell you a lot about how they see your consumption as a plant. Understanding the details of their own rates with their suppliers will inform you about the opportunities to lower your costs. If they purchase their energy from another entity, that is often found by mirroring the steps your utility takes to reduce their costs. You can create win-win scenarios.

What is the Cost/Danger of Not Considering the Details?

You get the level of services out of utilities based on the

level and the quality of the questions ask them. It is worth your while to craft very targeted questions for anyone you happen to be talking to. By phrasing your questions with precision, it will show them that you are serious and it will also allow them to get you exactly what you are after and not waste your time or theirs.

Going with the Grain is the Ultimate Frictionless Solution

While it would be nice to price shop your energy as you do with other things you buy, it just doesn't work that way. Your utility or commodity supplier is merely a conduit to the broader market. There are not huge deals to be had that will drastically reduce your bill on the supply side. Therefore, the battleground for cost savings will be most efficiently managed within utilities' existing rules.

Understanding your "relative risk" versus your absolute risk is a concept that must be wrestled to the ground if you want to maximize savings. We become accustomed to back-of-the-envelope thinking about this topic, and the thought of drilling down and understanding risk in any other way is foreign to us.

I want you to get this because it is the most powerful tool I am going to give you.

Relative risk is a term we are borrowing from statistics and medicine. It is used to measure the risk of intervention or change.

Figure 4-1. Risk Solution

To estimate the relative risk of some new action we are contemplating, we have first to attempt to understand thoroughly where we stand now and what our current risk is. We all have perceptions about what that risk is and those assumptions may or may not be correct.

We form those perceptions of risk based on our experience, but not necessarily based on experience related to energy. Instead, risk in general and whether having taken it on made our lives better or worse. Some people have a very negative association with even the hint of adding the most incremental of risks.

We must be objective when it comes to energy risk. What has happened in the past may have little bearing on the present or future.

In our study of risk in the world of energy, how precariously we stand regarding a potential interruption event, for example, is very important to assess and come to grips with. We like to think of our utilities as being able to supply us with a continuous supply of energy. However, utilities are human-made systems that get overstressed and break down. It happens all the time. I almost never talk with an industrial manager who does not experience temporary power brownouts, regardless of the level of firm power they buy.

Most of the interruptible power programs out there on the market are used to give the utility some relief if they find themselves in an emergency. These emergencies occur when generating assets are pushed to their limits, either during a summer period of extreme heat or a winter period of extreme cold. When this equipment is always running and is therefore difficult to maintain, it can't be taken down for preventive maintenance. What happens? The generating units break. When such a generator trips offline, your utility must react.

So, to analyze relative risk, first, we need to understand where you stand now. How much risk of tripping offline are you currently living with? Since your utility is required by ancillary agreements with other utilities to maintain the integrity of their system, they are required to do whatever it takes to keep *their*

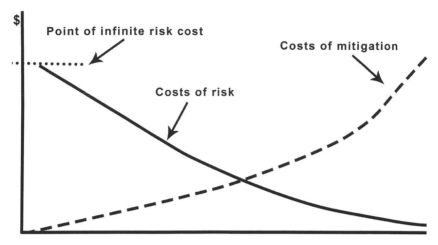

Figure 4-2. More on "Relative" Risk

emergency from turning into a rolling emergency for all adjacent power companies.

What does that mean for you as a manufacturer? It means that you are currently exposed to more risk of interruption in an emergency than you are likely aware of. Your utility can turn off your power at any time without telling you. That is if it is necessary for them to do so to maintain the integrity of their grid and thereby the integrity of interconnect agreements with neighboring utilities.

To precisely assess this risk, it may be helpful to understand the basic components of a statistical or medical risk assessment. To do that, we must dig down and find out both a philosophical reason and, if possible, a numerical representation of that incremental risk. To do that, we circle back to the task of learning a lot about how your utility operates.

We need to understand the likelihood of an emergency shutdown and then understand the internal mechanics of how the utility accomplishes such a shutdown.

Who decides to interrupt? What deliberation goes on internally before that decision is made? What other means does the utility provider try first to alleviate their emergencies before they call on your interruptible load to get them out of their emergency ditch? These are all questions that you are entitled to have answered and are usually accessible if you are willing to ask the right people the right questions.

To get this level of information, though, you are probably going to need to make a trip to the utility's control room and interview the people involved in the actual process. You will need to ask them to explain how things work.

Find out, too, what is the worst thing that could happen with your current level of firm or interruptible service. And you need to understand their projections of demand and supply for the future. For example, the EPA is demanding the retirement of coal generation and that event will influence demand and supply. It is essential that you get a feel for where the utility is spending money to become more reliable.

The factor that is best utilized to get a grip on understanding the utility's future is their predicted "reserve margin." That is the amount of "reserve" power supply capability that the utility predicts they will have over and above what their requirements to serve their customers is predicted to be.

Predicting something like this is complicated. How many people and companies will move into and stay in their service territory is not a simple thing to know. It involves understanding local demographics as well as understanding how aging infrastructure can be utilized and upgraded to serve those needs.

As we have discussed, utilities err on the side of being conservative. Their predictions for reserve margin will likely be tight. For our purposes, that means we can have some confidence in those numbers.

Every utility has a different predicted reserve margin that goes out many years in the future—that is likely different than their actual reserve margin, and that is information you should be able to learn if you ask the right person and are tenacious about finding out.

These are the questions that you need to ask to assess your relative risk of being interrupted. But, don't feel that you need to understand all these terms before you dive in and start asking questions. My recommendation is, do not try to become a utility expert but focus on getting answers to questions about reliability using your own terms.

Once you understand where you stand now, you can begin to understand the likelihood of extraneous events like emergency power curtailments. Since we know utility representatives are by nature very risk averse, they will likely interpret the data very differently than you would as a businessperson or manufacturing manager.

Again, the questions are straightforward. What would have to happen at the utility's control center for an emergency curtailment to be called? Who would be involved and what criteria would they use? What is the protocol the utility uses to resolve emergencies? Does the utility have sufficient backup generation to avert prolonged weather events? Have there been power generator emergencies in the recent past and how were they resolved? What has changed about the power delivery system to make power interruptions likely now or in the foreseeable future?

When the utility has called for cutbacks in power consumption for interruptible service customers, was this action the last straw or the first resource the utility used? Why did they call for the interruption?

Also, you want the utility to know that all this matters to you as their customer. If they have no contact and no feedback from customers regarding the hassle factor their actions put you through, they tend to look at a power interruption like no one is negatively affected. If they are told about what you went through to meet their power curtailment and made to understand that people's lives are turned upside down, that product losses occur and the like, the utility is a lot more likely to use other means to resolve their power emergency problems next time.

By now, you see why a thorough and objective risk assessment is essential. Although I may look at power interruptions as a hassle and a lot of trouble for a factory to comply with, that isn't the case with all customers.

It is a fact that some manufacturing plants enter interruptible power contracts without understanding any of these concepts. They don't mind the interruptions. Why? Because they simply shift operations into preventative maintenance mode when an interruption takes place. These plants do not look at the interruptions as a calamity to be avoided but as an advantage. They can accomplish necessary maintenance and get paid for it through the interruptible rate capacity and energy payments. The dollar savings within interruptible power can be enormous, and often the money received can be enough to ensure a manufacturer's profitability for the quarter or even the entire year.

Another conclusion a thorough risk analysis may lead you to is to mitigate your exposure by adding on backup generation. If the utility's system seems unstable, you may benefit in several ways from spending money on backup generation. Having a backup generator at your facility can protect you from the power company's lack of redundancy by offering you access to a continuous source of power in the case of the utility's failure to provide reliable power.

Some industries like data centers require a backup system, regardless of how reliable the utility seems to be. The downside risk to those businesses that depend on uninterrupted power for the utility's failure is just too much, so they invest in 100% backup generation. (Any company that has backed up their facility with 100% generation and does not opt for interruptible power from their utility is making a big mistake, but it is quite common.)

It is possible now to purchase natural gas backup generation for your facility or even rent it. With a little advanced planning, rental units can be brought in with very little advanced notice and parallel with your existing power delivery infrastructure or substation. It is possible to pay for an on-call relationship with your utility provider and only call in and pay for the generation when you need it too. Batteries, like those manufactured by Tesla, are increasingly an option as backup systems.

Another critical element of risk assessment is understanding what happens at your facility when the power goes out. You may be fortunate enough never to experience momentary power sags that trip your systems and force all your sensitive electronics to restart. That is not the norm though. Most large facilities have gotten very used to rebooting their operations after a power blip.

So, take the time to understand your company's worst-case scenario. What losses would you suffer if you had to contend with an extended power outage? Is there a way to maintain the integrity of the material that is in process without having to throw it out? I once worked with a smelter that found a way to seal their molten vats of metal during a power interruption and keep it salvageable for 12 hours. They saved enormously by making their operation interruptible.

I know interrupting operations is not the ideal scenario, since you are in business to make things. But if you had to, what could you do during a sustained power outage if there were significant savings in the balance? You must be willing to stretch the operation a bit to save significant costs in the unlikely event of an extreme power emergency.

Keep in mind again that your utility would not think twice about taking your power offline if it meant maintaining the integrity of their system.

By coming to grips with what could happen to your power supply, you are in a lot better position to assess how you need to react to those potential eventualities as a business. It is highly likely that an investment in backup generation would merely be a nice to have and not a necessity.

To circle back around to the beginning of our discussion, it is essential that you make a detailed survey of all the potentially relevant facts regarding the stability and reliability of your current situation before you can properly assess any alternatives. You also need to be bold and err on the side of being aggressive if you want to save money. It is likely you are far more exposed than you think you are now. The possibility of getting interrupted on most programs is relatively low. The interruptible power program you are reflecting on is probably a winner for you if you can be creative.

RELIABILITY IS IN THE EYE OF THE BEHOLDER

Whether a system is reliable or not is a subjective call. One man's reliability is another man's risk. The world is replete with stories of over-engineered systems that failed. I just read a story about a DC-10 airliner that crashed in Iowa back in the 1980s. The airplane had a total failure of all control surfaces. The only thing the 4-man crew could do to manipulate the movement of the aircraft was to differentially switch the thrust levers to get some skids to the left and the right.

The reason I am telling you this is so that you can put risk and reward in perspective. The DC-10 had redundant control systems. And when I say redundant, I mean 3X redundant. First, hydraulics to back up mechanical systems and then also a back-up pop-up system to drive the hydraulic pump if that system failed in flight. Remarkably, most people survived the inevitable crash, but only because of the phenomenal luck of the crew.

Although we can reduce the chances of error and risk, we cannot eliminate them. Why? Because it is not just the closed system or product like the aircraft that influences the likelihood of a catastrophic failure; it is outside factors, like in the case of the airplane, weather, and a breakdown of the turbines. There are just some contingencies that are so unlikely that they are not worth realistically considering as a part of design of the system or process.

When it comes to analyzing the risk in power delivery systems, it is similarly impossible to understand and correctly weigh all the things that could happen to interrupt you. Could a terrorist crash his rented Lear Jet into your substation and knock you out of business? Yes. Is it worth trying to protect yourself from this, or is that just one of the many possible but unlikely events that your business is exposed to?

Reliability and its definition for you is organic and ever-shifting. It is different than anyone else on the planet's assessment of that very same risk. You can fully understand a system and turn over every rock and still not fully understand what could happen in a worst-case scenario.

There is very little that is free in life. You know that, but it is easy to forget. Increased income or success in any way usually comes at the price of increased exposure in some way. In many cases and I would argue that in most cases with utilities and energy, the rewards far outweigh the risks. That is because you are already "interruptible," and you just don't realize it. Your plant is subject to being interrupted by the utility for its own reasons or by the weather or by a random act. Things happen. We fool ourselves to think that our "firm" power is guaranteed.

So, after you have analyzed all the options and all the reasons an interruption could happen, you need to mix in some healthy reflection on how much reliability you *really* do have now. The systems we rely so much on were designed by humans and are influenced by human interaction as well as by random forces of nature and civilization.

So, what does this perspective tell us to do? It tells us to be more aggressive in our quest to save money in energy and to use risk-shifting to reduce our costs—not to play things "safe."

How Do You Know When You Have Analyzed Your Risks in Sufficient Detail?

Because it is "risk" we are talking about, you could justify analyzing your participation in perceived increased levels of it *ad infinitum*. For every person, there is a different level of analysis, and none is precisely correct. That is because you might be comfortable with a warm and fuzzy feeling at the concept level and I might be the type who needs to back up that gut feeling with a lot of data. We also may work for companies that require 3-4 quotes to buy a pencil sharpener. If that is the case, then you will probably make your decision to take on more risk exposure at the emotional level, but you will need to articulate that comfort level with a lot harder facts.

Always make sure you have a plan B and maybe a plan C.

Doesn't Business Interruption Insurance Cover Voluntary Acts Like Demand Response?

Unfortunately, not. Business interruption insurance is meant only to cover natural events like Acts of God. It is not intended to cover premedicated savings opportunities like this.

The Upside to Downtime Is It Keeps You Up and Running

There is nothing more frustrating than having your equipment fail right in the middle of a big order. Most every manufacturing plant I have seen keeps its equipment up and running by averting maintenance disasters.

These planned periods of downtime are, therefore, a necessary part of the overall process. Customers understand machines that have high up time must be taken offline every so often to keep them running.

The Downside to Downtime Is You Aren't Working, but That is Only Part of It

Of course, the downside is your plant is no longer producing widgets if you are down for any reason.

Even though you are not producing your product during "downtime," it is time you can use to catch up on other things that a plant needs to do to keep going. This is the way that downtime events should be viewed. Sure, you are not making your product, but these events give you time to sharpen your saw.

Which Perspective is Correct?

Both perspectives have their ring of truth. While you should be more ok than you are about "suffering through" downtimes, you don't want to engage in new ventures that will make those downtime events so frequent that you can't meet your customer orders. This is why it is so important that you carefully analyze interruptible opportunities from all angles.

Aren't I Being Irresponsible Exposing My Business to Power Interruptions?

You are the most responsible as a business owner or manager when you continually search for the very lowest price possible to pay for the energy you use. For manufacturers, that means exposing the risk and reward continuum to the light of day. Most people stay stuck at a high level and never really examine this risk versus reward continuum in any detail. Instead, they use a summary judgment approach and only examine things casually. No one wants additional risk of any kind at first glance.

That is our nature as humans. We want all the risk factored out of all our situations in life. The utility is more than happy to absorb that perceived risk for us for a price. You are likely paying a substantial premium for your power and gas if you buy it firm.

If we aren't careful, risk aversion can become our *modus operandi*. But, this is where the opportunities in life lie. Do we need power and gas that are always going to be available when we ask, or are we ok with living with a few downtime events a year that we can use to do preventive maintenance that will have to be done anyway? Can we shut off peripheral equipment and keep running for the most part and still produce our products? Only a thorough examination of all the relevant information will yield the kind of answers we need to make these calls in our businesses.

The funny thing about risk and the concept of it is how we are surrounded by it every day and do not really pay much attention to it or qualify it. It is only when we *perceive* that we are taking on more risk that we start getting nervous. I suppose that is why the Evil Knievels among us still get shaky when they try something new.

CHAPTER SUMMARY

Understand the relative risk that alternatives for cost reduction present. To know the risk of other options, you must first know and assess your current risk factors. Often you can save a lot of money by taking on a tiny additional risks.

QUESTIONS TO THINK ABOUT

1.   Under what circumstances can the utility interrupt my facility now without notice?

2.   What are my real costs in downtime and waste associated with random utility interruptions?

3.   Is there a record being kept of when and for how long these power interruptions occur?

4.   Does my electric utility sometimes interrupt my power randomly due to infrastructure problems (like animals shorting out their delivery system)?

5.   What is the likelihood I will get inadvertently interrupted without compensation? (For most utilities it is in the 80% range.)

Chapter 5

Real Time Pricing (RTP)

Our strategies so far have been about shifting theoretical risk from the utility to your company. Or, at least shifting perceived risk. Real time pricing is no different. With RTP programs, your monthly bill will be calculated based on what the utility's costs are at each hour of the day.

Right now, if you are on a general service or time of use rate, you are paying a lot of "insurance" in the form of averaged pricing. All the peaks and valleys and variations have been eliminated for you. You pay a flat rate, depending on what block of time you consume that energy. The price is set. You know what you are going to pay in advance and it doesn't change very often. But it is a lot extra that you are paying.

With Real Time Pricing, you pay the utility's actual cost of generating your power plus an adder. In an RTP program, you will pay higher prices in the summer. In the winter the prices will be much lower than you are paying now with a general service type of rate.

The RTP rate class owes its existence to utility competition for new customers. That is, utilities competing to entice manufacturers who want to locate in your state. Utilities must also offer to you the pricing they have worked out for those customers.

You will always come out better paying for energy this way; especially if your operation has time flexibility. You can buy your power very inexpensively at times. If you can self-generate and thereby make your own power sometimes when the RTP price is very high, these rates can save you even more money.

With generalized rates, you are not exposed to the ups and downs of the hourly market for power. During the summer, those prices can fluctuate by over 200%.

It is not uncommon for large accounts to save 10% without making any changes at all to their operations.

MORE ON REAL-TIME PRICING

How Much Advanced Notice Do You Get on Actual Prices?

As we have said, the price you pay for electricity will vary hour by hour based on actual market prices. RTP often uses "day-ahead" prices, meaning the price for each hour of the day is set the night before. These are your utility's best prediction of what prices will be from late in the afternoon the day before the energy is consumed.

Your bill is then computed using those hourly, day-ahead market prices and the amount of energy you use during those time blocks.

Summer Pricing Patterns

Electricity prices typically rise the highest during the summer months of June to September, depending on the part of the country you are in.

Those prices are the highest during the late afternoon and usually peak around 5 p.m. when people get home from work. So, you will find prices correlate well with weather intensity. On weekends and holidays when people are at the lake and industrial facilities and commercial office buildings are not running full force, the prices are usually a lot lower too.

Seasons Other Than Summer

Non-summer prices are considered off-peak and much lower than summer prices. As well, the pattern of low prices is usually a lot more predictable and lower. Heat is supplied in the winter from natural gas systems by and large, so there is less demand on the resources that produce electricity during the non-summer time. During the non-summer months, you really don't even have to consider cutting back on power based on the price because the price you are paying will be so much lower than the average amount you had been paying on your previous general service rate.

Weather Impact

Weather is the critical factor driving electricity prices. Weather and its effect on the expectations our society has for climate control drives most all energy systems. Even in the winter, the price of natural gas heating requirements can affect the price of electricity. Why? Because more and more, electric utilities use natural gas-powered generators to make power. Extreme cold can drive up the need for natural gas for heating in the winter and thereby drive up the spot market price of gas.

Check Prices When You Need To

While we have been talking about day-ahead power markets, there is also another version of RTP that can reduce your costs even more—hour-ahead pricing. With hour-ahead pricing, the system is very dynamic.

In this system, you get your information about the cost of the power you will consume only an hour ahead of when you use it. Hourly pricing has a smaller "adder," and thus, it can be much cheaper than days ahead. Whether you react to those fluctuating prices and reduce load is altogether a different matter. Most utilities will be transparent with you about their costs to generate power once you enter an RTP contract. These data are called the utility's "system lambda" or their cost to produce the next hour of power. These numbers are continually electronically filed with FERC in Washington.

In general, to qualify for RTP rates, you must be increasing the size of your operation.

When your utility agrees to serve your plant, they are also deciding to invest in your future at that location by installing step down voltage transformation and the wires needed to carry the power you need to your buildings. Because of that, utilities can recover that investment from ratepayers over time. You as the utility customer are usually allowed a certain amount of investment, and then anything over and above that you will have to pay for yourself in the form of an "extra facilities charge."

The only way to find out for sure if you will save is to ask your utility to simulate your savings on the RTP rate. Your utility has all your metered data going back several years so you can ask them to run a comparison between what you are paying now and the RTP rate option.

If RTP pricing shows a lower cost for you in the simulation, you must then establish what most utilities call a baseline of power use. You will pay more for that baseline part on a generalized rate. That baseline will be priced most likely at your existing rate so your utility can continue to recoup their initial investment. So, part of your power will be priced at the old higher rate, and the rest of your energy use will be priced at the much lower RTP price. I have seen RTP priced power cost roughly half of the traditional generalized power pricing. Not a bad deal for taking on and living through the frequent price swings.

These are all ways of shifting a little of the risk the utility has traditionally taken on for you through their generalized rates. Real-time pricing programs are a way for the utility to pass on savings to their customers who understand energy pricing.

How Does This Apply to Real-Time Pricing?

RTP is like any other project you analyze. Proving this concept to yourself is impossible without wrestling with the details. Everything in life boils down to getting comfortable with those details.

Could RTP backfire on you? Could it end up costing you more than a general service rate costs now?

It could. In the short term only, however. In the long run, no way. How can I be so confident in that? Because actual costs ultimately drive all utility rates. The costs that you would pay today under an RTP rate are the prices everyone else will pay in subsequent years as those actual costs to produce power are distributed to the rate base of customers.

Like in the insurance industry, you are in effect becoming the "underwriter" for your company on the energy issue. Peeling back this veneer and then understanding the real nature of the risk you are dealing with is the key to reducing costs. You get to see the real nature of the risk you are exposed to and make a risk/reward decision about how much additional risk you can expose yourself to and still operate.

CHAPTER SUMMARY

Real-time pricing can save you a bundle of money, especially if you reduce consumption when the cost is too high. You get paid for being flexible with your utility. There are two versions of RTP pricing: day-ahead and hour-ahead. You must be the most flexible when you get your energy prices only one hour ahead, and that program is the most lucrative.

QUESTIONS TO THINK ABOUT

1.   Does my electric utility offer real time pricing of power?

2.   Have I asked my utility to simulate my last 12 months usage priced on their real time pricing program?

3.   Do I have the ability to curtail demand if real time prices spike?

4.   Are other industries like mine taking advantage of a real time pricing program?

5.   Will my utility give me visibility to their hourly prices so I can choose to not buy if prices are too high?

Chapter 6

Coincident Peak Costs

WHAT IS IT?

You will often see it referred to as 4CP (or even 5CP). 4CP means "Four summer months' coincident peak."

During the four summer months, the power supply is at a premium. If you can reduce demand during the 15-minute intervals when your utility hits its own peak, then you can save a bundle of cash on demand costs the rest of the year.

The reason these programs exist is many utilities have similar agreements with their generators further upstream. A critical aspect of many of their generation power agreements is a provision that the utility will be charged a demand charge during the peak "coincident period." Coincident is your demand when the utility or their supplier is reaching its peak for the year. This is a case then of your utility helping you save money if you help them save money.

HOW IT WORKS

In this case, your utility would be purchasing its power from a 3rd party generator. That is easy to find out—you just ask them, "What is the nature of your agreement with your supplier? Do you pay demand charges based on your peak demand during the summer? If not, what determines the amount you pay for power?"

This is the ultimate win-win way to save. Your company helps your utility shape its load profile, and they pass those savings on to you. If you are helping them achieve a lower overall demand for their system, it benefits everyone.

If your utility does not have a formal coincident peak program, that is ok; you can save anyway by drilling down and understanding how your utility is billed by others. If you know the details of how your utility is billed by their suppliers, you can potentially create your own program. Just ask your utility if they would be willing to pass along any savings you achieve for them by reducing demand.

Ask, and Ye Shall Receive

Savings do not just jump out and demand that you take them. They must be sought out, and they require an attitude of creativity and insistence. That means that you don't give up easily at the first sign of resistance.

While these savings are achievable somehow with most any utility, it will take some work to uncover the potential.

Utilities are used to operating within the bounds of simple and straightforward programs. Utilities don't get asked many questions like this. But the way to save money is to ask these uncomfortable questions.

Utilities many times like to act as if they are an inflexible government entity. That attitude harkens back to a day not too long ago when utilities were just that—real government entities. Many of the attitudes we encounter are holdovers from those days. They didn't have to offer anything since you, as a consumer or business, were a captive market—a market so captive that you had no choice but to buy from them at the price they selected.

That is where most of us got our belief—wrongly—that we should just pay our utility bills and keep quiet.

As you know by now, nothing could be further from the truth. There are millions of dollars hidden between the lines of your bills and in utility contracts.

Just like the interruptible or demand response programs we have talked about, there are things you can do to make the most of any downtime created by a load peak reduction event. So, these interruption events don't have to be all negative.

For example, every business performs preventive maintenance on its equipment. During this maintenance phase, the equipment, of course, must be out of commission. If you are called on by your utility to curtail in response to a grid emergency or a coincident peak reduction event, why not just go ahead and do your preventive maintenance then, rather than wait till later? If it must be done anyway, just do the work during the event. Not only will you make your equipment more reliable but you will be getting paid for it through the CP program.

Do not be discouraged if your utility does not roll out the red carpet and give you a trophy for asking the question. Utilities may view outside-the-box questions as a lot of work for them.

The CP plan will need to be interrupting your power 4-5 times per month for a few hours during the summer to catch the actual 15-minute peak for the month. If you had a crystal ball and knew in advance when the utility peak would occur for the month, you could curtail only once during that time and capture the exact 15-minute interval you need to. It is a lot more likely that you will miss if you try too hard to optimize.

Instead, it is best to curtail along with the utility on coincident peak reduction if you can. Utilities stand to lose the most from missing the peak, so they may be liberal in curtailing load on any day and time when they think there is a good chance the peak will occur.

Don't even think about trying to figure out the peak on your own. If you would like to reduce your number of curtailments, there are numerous services out there you can pay that have more refined analytics for predicting peaks. With years of actual data with actual interruptions in their databases, these services are experts at predicting peaks. While they won't nail it down to one interruption, they can reduce the 4-5 times a month called by the utility to 2-3/month.

WHY COINCIDENT PEAK PROGRAMS
CAN BE SO POWERFUL FOR YOU

It is a win-win for all involved; not only do you win by having lower costs, but the utility wins by not having to pay their supplier extra and not meeting peak demand charges that are built into their contracts. The supplier of the utility also wins because their costs are down as well by acknowledging everyone's ultimate incentive—which is to keep costs down—and making sure that everyone understands them thoroughly, all involved can work together to minimize costs. Everyone is working together to make the most of the arrangement that the utility has with that supplier. The system works very well because of that.

4 CP Also Helps the Environment

Coincident peak programs help reduce peak demand at critical times. That means that stakeholders throughout the system are minimizing their maximum power consumption at times when the system is in overload mode. Electrical delivery systems were designed many years ago to compensate for the maximum electrical demand that could occur at any time of the year. This system has been found to benefit the utility but not everyone else. What I mean by that is the utility makes its money by adding costs—the more infrastructure and expense they can add, the more money they can petition the public utility commissions to pay them a return on. We live in a world where the immediate availability of anything we want is expected. This availability comes at a super high cost. However, we cannot afford such a system any longer when it comes to power delivery. If every time a new power peak was set, the utility searched for a way to supply the power, the system would be so overloaded that generation facilities would be everywhere.

So, coincident peak systems do just that. If you are willing to play along and reduce your power consumption several times during a given summer month, the bottom line is you will be able to cut your demand charges substantially. Knowing that demand charges could be as much if not more than 50% of your entire bill is a strong incentive to get creative and find a way to make those demand reductions happen. Knowing also that when you reduce demand, you are helping the broader society at large keep from having to be inundated by new power generation facilities, is also an incentive.

Think about it this way: Taking part in demand reduction and coincident peak programs is a way that you can, as a business, make a positive impact on the entire power delivery system.

CHAPTER SUMMARY

Coincident peak reduction or four-summer month coincident peak is a grid-level program that rewards your company for mirroring your utility's summer demand reduction patterns. You must reduce demand a few times in the summer months to save money on coincident peak. By carefully following your utility's recommendations on load reduction, you can reduce your demand by up to half for the next 11 months.

QUESTIONS TO THINK ABOUT

1. Have I asked my electric utility if coincident peak reduction pricing is available for me?

2. How much annual benefit could I receive for participating?

3. Could my facility perform preventative maintenance during curtailment periods?

4. Does the utility offer a coincident peak reduction event notification service?

5. Are there other event notification services that had success predicting the monthly peak more accurately than my utility?

Chapter 7

Solar Power

WHAT DOES IT ALL MEAN FOR MANUFACTURING PLANTS?

Right now, most energy discussion surrounds how we can most efficiently deliver fossil fuel derived electricity made in faraway plants to our homes, factories, and offices with the least impact on society and the planet. There are enormous losses involved in our distributed power model of transporting electricity long distances over wires. A lot more power than is needed must be generated just to transport the power to you.

There is also an aesthetic price we pay for the current model as well. Electrical distribution wiring mars the beauty of our landscapes. In some places, utilities have been burying their cables underground, but in many more, the wires are installed overhead, so dense they almost block out the sun.

In the future though, there will be less and less need to string wires along the streets, as power will be made near you instead of far away.

WHAT IS SOLAR EXACTLY, AND HOW DOES IT WORK?

Solar power has become ubiquitous in some areas, but still absent in others. Solar power means capturing the energy the sun deposits every day on the earth. Approximately 1,000 watts per square meter. There is enough solar power hitting the earth every day to supply all our power needs. As prices for solar power generating systems drop, this is becoming not only a way for manufacturers to take back control of their power costs, but also to raise the worldwide standard of living, since the majority of the countries that need power are those with the most sun. One problem with solar, as with wind and wave energy is that the power, once harnessed, cannot be stored for very long. It must be saved, used, or lost forever. Therefore, there can be no meaningful discussion of solar power without discussing battery power at the same time.

Industrial-sized batteries are being developed that can be charged by solar power generating systems and serve as backup power in case of power interruption. That means as battery prices drop we are just millimeters away from being able to power an entire industrial facility with solar backed up by batteries. Banks of batteries the size of shipping containers can be brought to an industrial facility and can provide 2-4 hours of backup energy.

How Much Does Installing Solar Cost?

The cost to manufacture solar panels is dropping quickly. The good news is, once you have installed your solar panels and peripheral hardware, there are minimal ongoing additional costs, other than cleaning them periodically. That, of course, means that the price per kWh is susceptible to the "first cost" of the unit you buy.

Any number I were to give you now will be much lower by the time you read this. In early 2016, the average price was $.122 per kWh. The good thing about this is that the price is irrespective of location—the cost to produce solar energy is about the same everywhere, although the total credits and rebates could be dramatically different. However, regarding the justification of expense, it matters a lot where in the country you install the solar panels because the price of energy varies so much in comparison.

While the current price of energy to compare solar within southern states is relatively low-priced ($.05/kWh), the northeastern states, as well as the west coast states, can be 2-5 times that much. So, it doesn't make as much economic sense to buy solar energy in the South, but with government subsidies and falling first-prices, it has become the obvious thing to do in increasingly more places.

Since many coal-generation plants are becoming too expensive to repair, the prices for utility-supplied power in the South will increase. As those costs rise, solar will become more and more attractive there. As well, in the meantime, the proliferation of solar energy in the Northeast, where power prices are very high now, will serve to bring on more adopters and drop the first cost of solar for everyone. This cycle of utility price increases and greater adoption of solar through government subsidies will perpetuate improvement.

Why Buy Solar Now vs. Wait Until Prices Come Down?

There is always an opportunity cost for not taking action. With opportunity cost, we are measuring the incremental benefit of taking action over and above what we are doing now. Whether it is a solar investment we are talking about or something else, it is in our best interests to carefully measure the often-overlooked costs associated with doing nothing—this "base case" scenario of keeping things the way they are.

Most people look at opportunities in isolation, however. That thinking seldom gets challenged even by the equipment vendors. However, think carefully about the actual costs of keeping things the same. What I mean by that is, the purpose of solar is to reduce your electricity cost per kilowatt hour. If you take no action and never install solar, you would continue to pay ever-rising electricity prices provided by your utility. Utility price inflation is 4-5% per year. Acting on solar or any system that disengages you from the grid at least partially will forever displace that energy that you would typically be buying at those ever-increasing prices.

Prices come down; technology improves. The sooner you tap into the limitless source of free energy solar provides, the better. Solar energy can be monetized in two ways: selling the energy back to the utility or using it yourself. The buyback of the energy you make from your utility will be at approximately what it costs you to buy the energy from them. It isn't important how much you think the energy is worth or how much you had to pay to set up the system to convert the sun's energy to usable energy. Utilities determine that value and are only willing to pay you the amount it would have cost them to make those same kWhs for you in their fossil fuel power plant.

Once you have paid for your solar project through selling energy back to your utility or consuming it yourself in your operation, your mindset needs to be that you now have a source of incremental free energy. That means energy created by the solar panels is essentially free since there are only minimal costs associated with maintaining the system after initial payback of the investment. All the energy you make becomes a gift.

Rather than getting hung up on the payback period, I would encourage you to look at it this way—passive energy created via solar gives you the opportunity to supply yourself with a perpetual source of power without very much maintenance cost. Producing solar energy is like having an ATM stuck open.

How to Decide which Solar Equipment to Buy?

In answering this question, consider shopping for a vendor as much as for your specific solar power system. All the vendors will, of course, seem knowledgeable and convincing.

One of the central questions you need to ask yourself is what are the lease and the purchase options available? Know up front how simple the vendors you consider can make getting started.

Another consideration when shopping for a vendor for your solar power system is knowing the *total* cost of the system.

There will be a wide array of answers to that question. You will find it helpful to create your own spreadsheet that includes options offered by different solar power vendors. There are a tremendous number of solar panel manufacturers and consultants. You need to do your homework and vet the supplier and the technology. Although there will be several options available in the future that integrate solar into existing building structures (like coatings) that do not rely on building real estate to function, the current primary concern is where you will put the solar power equipment. Designs like Tesla's will eventually abound that use standard construction components, like roofing materials, to integrate solar into the actual structure of a building (and not require any special designation of roof real estate or ground real estate for installation of the solar panel system).

It is the equipment you can afford to purchase now that is the best equipment to own. It might not be worth the trouble to consider options for leasing the equipment because much of the credit that you would receive because of owning the system will be lost to you and gained by the lessor.

How Can I Make a Solar Power System Earn Its Keep?

The first stop is the federal solar tax credit. The solar tax credit (or, more appropriately called, the Investment Tax Credit). This is not a reduction in your tax basis or the amount of income your income tax is calculated based on; rather, it is a deduction from the amount of the actual tax that you pay.

The U.S. government likes to incentivize investment in new and emerging technologies that are difficult to justify because of the initial expense.

New products on the market, of course, carry an inordinate amount of research and development costs. Only the intrepid seriously consider technology right out of the starting gate for this reason. Without some buyers opting to use the technology, the costs are too high for most to give it a try.

The role of government is to absorb some of the risks, since the eventual adoption of the technology provides such a vast benefit to society at large. Solar falls into this category.

With any new technology, only the passage of time can make it evident to the masses. No one wanted cell phones in the beginning and didn't see the need for them. Even the wristwatch when it first appeared was dismissed as jewelry a man would never wear. Everything that now seems obvious and impossible to do without, at one time, seemed like a dumb idea. Someone or some entity had to underwrite the idea until it took hold and became indispensable. Someone or some entity had to believe in the concept so firmly that they were willing to withstand an extended time of no economic return before the audience, and the market was developed, and interest reached critical mass.

It is seldom a comment on the viability of such a product that it takes a herculean effort to amass an audience. It is just the way society adopts something new. Eventually, the product or service is accepted by the public at large, and it is finally called the obvious solution and the right thing to do.

What Parts of the Solar Investment Are Creditable against Taxes Owed?

The entire investment should be treated as a credit. You will have to have verifiable documentation from a third-party installer who will provide the documentation that the federal tax credit requires. Most, if not all, contractors in the space will be able to give documentation because the ability to document their work is the essential ingredient of their success. Solar vendors must be experts at understanding and applying federal state and local tax credits.

You may qualify for additional benefits provided by the government. For example, the food and drug administration offers grants depending on the size of the company and the type of industry that you are in. It is worth looking into this further. I was recently involved in a project where nearly 1/3 of the cost was funded by such a grant. It took a long time and a lot of trouble and loads and loads of government paperwork to get the credit, but it was successful.

The University of North Carolina supplies a lot of information about credits and rebates in a database called DSIRE. Inside this database, you will find resources for renewable energy options in every state. Not only at the state and federal level but also at the local level where cities and counties are incentivizing companies to invest in solar and other renewable technologies. Going to the DSIRE database can save you a lot of time and energy. If you are confused about any aspect of renewable energy or solar power, DSIRE will answer your questions and they are only a phone call away. DSIRE is an excellent service and does not cost a dime. I encourage you to take full advantage.

It is always a good idea not to assume vendors have all the credits and grants at their fingertips. For example, in the lighting projects I just referenced, three of the four vendor options had no knowledge regarding the federal grant. They knew about the tax credits, but they were not familiar with the grants. Do not settle for what the vendor brings you, even though the vender seems the most incentivized to get it right. That does not mean they know everything.

As the saying goes, if it were easy, everybody would be doing it. With all conservation projects, you need to be your own advocate. Many vendors want to incentivize you enough to act, and that is all. If you are happy with a federal tax credit, then they feel they have done their jobs, and they do not need to dig deeper for more credits to incentivize you to purchase their unit. (Tax credits, by the way, provide a dollar-for-dollar reduction of your income tax liability. This means that a $1,000 tax credit saves you $1,000 in taxes. On the other hand, tax deductions lower your taxable income, and they are equal to the percentage of your marginal tax bracket.)

What Guarantees Are Available with Solar Systems?
With most installations, solar panels will carry guarantees and warranties. The first is a performance guarantee and the second is a warranty on the equipment. What is most common is for the performance guarantee to be set at 90% for ten years and 80% at twenty-five years.

What Problems Can Solar Panel
Warranty Insurance Prevent?
If you install a photovoltaic system known as a PV system or solar PV system for your home or business, by far the most likely scenario is that your panels will perform trouble-free for decades. But, solar panel problems do occur from time to time. If you en-

counter a problem with a solar panel after installation, it's critical for you to understand that one problem will usually not render your entire system inoperable.

Depending on the nature of the panel failure, the rest of your system will continue running with the failed panel left in place. But occasionally, panels fail in ways that impact the performance of neighboring panels. In either case, it's the solar panel's product warranty insurance that will cover you if you ever need to exchange a faulty panel for a new one.

Evaluating and comparing the product warranty coverage of panel manufacturers can help assure you that your service and support needs will be covered if a solar panel problem ever occurs. Manufacturers' product warranties are therefore an important complement to other considerations in assessing a panel manufacturer's technical specifications, but also in assessing its business practices. In case a solar panel fails, the manufacturer will ship you a replacement panel and pay for shipping cost and labor cost to replace it.

How Many Days of Sunlight Are Required to Make a Solar Power System Pay for Itself?

All hours of sunlight are not equal but even cloudy days offer free power.

A "peak sun-hour" is an hour of sunlight that offers 1,000 watts of power per square meter, and this reflects intensity. These hours occur when the sun is highest in the sky—around noon, and they also increase in the summer.

The top 5 states with the most sunlight include Arizona, California, Colorado, Florida, and Massachusetts.

The aesthetic aspect of solar technology has kept many out of this game, regardless of the availability of sunlight in a state. However, Tesla and others are working to integrate solar and close to invisible.

Consider these recent innovations announced by SolarCity, Tesla's solar company:

SOLAR ROOF FROM TESLA

Last year, Elon Musk announced that SolarCity had developed high-design solar roof tiles as an aesthetic option to chunky solar panels. These tiles would tie into the company's wall-mounted battery during times of low sunlight. The tiles mimic the look of French slates, Tuscan shingles, conventional black asphalt shingles or curved clay shingles, offering a mix of smooth and textured surfaces.

The tiles will consist of a photovoltaic substrate covered by glass. Musk also claims that Solar Roof tiles will provide better insulation, last longer, and have an installed cost less than that for a normal roof plus electricity costs. Though some of these claims can be confirmed once the tiles enter the market, they do offer the look and feel of high-end roofing products. The resemblance to slate and terracotta offers a premium appeal, which makes you wonder how much they may finally retail for when they go on sale.

There is still no clarity about the actual costs of Solar Roof. Musk has said that it will be cheaper than regular roofing, though even approximate numbers haven't yet been provided.

Transparent Photovoltaic Cells

SolarWindow Technologies has come out with a revolutionary transparent solar panel that the company claims transforms windows into photovoltaics. When applied to a 50-story building, the panels can produce up to 50 times the electrical energy of traditional rooftop panels. You don't have to replace the windows of your home entirely, only treat them with a proprietary process.

As part of the process, liquid coatings are applied to glass and plastic coatings at ambient pressure, and the coatings are then left to dry at a low temperature to develop transparent films. This is repeated, and together, the coatings, glass and plastic surfaces produce electricity.

In 2014, scientists from Michigan State University showcased a fully transparent solar concentrator through which visible light can pass, and whose organic molecules move infrared and ultraviolet wavelengths to the edges of the glass. Here, tiny photovoltaic cells convert these radiations to electricity. However, more work is needed on the prototype, which has a conversion efficiency of 1% versus the 20%-25% efficiency of conventional solar panels.

New Designs in Solar Energy Systems

Solar energy companies in other countries have launched sleek heating systems, some of which apply a unique mechanism to heat air and water. SolTech Energy's tiles are made of glass and installed atop black fabric that absorbs sunlight. The air slots under the fabric heat clean air, and the warm air heats up the water. The tiles weigh the same as typical roof tiles and generate 350kWh of heat per square meter. Pythagoras Solar has come up with a photovoltaic glass unit (PVGU), a transparent solar panel that uses monocrystalline PV cells for generating power for buildings. The photovoltaic strips collect the sunlight, while visible light diffuses through the windows. These solar windows have been installed on Chicago's Sears Tower.

Monier's solar tiles integrate with existing roof tiles without the need to be bolted on top of traditional roof tiles. The tiles strike a balance between preserving the look of buildings and helping achieve energy savings. This energy system offers a power output of 120 Wp per square meter. Some, like Tractile, use interlocking tiles that are quite light-weight compared to concrete roof tiles.

Thinking Beyond Beauty

Solar panels are an energy-efficient solution for not just Fortune 500 companies but also small and medium-sized businesses, which have the opportunity to make the most of the financial incentives available for solar installations. According to EnergySage, businesses with solar power systems offset their energy usage by 75%, amounting to monthly savings of slightly over $1,400. Businesses break even on their solar panel systems within three to seven years, and the system itself can last them anywhere from 25 to 35 years.

Over the years, the costs of installing solar panels have fallen, and with electricity prices remaining quite high, the move to solar does make practical sense. There is another cost benefit : you can deduct 85% of the cost of the system from your taxes, which will help you further save money on the purchase cost of a system array. While some states have solar renewable energy certificates, New York offers the NY-Sun

Incentive Program that provides financial incentives aimed at lowering the installation costs related to solar electric systems. The industrial and commercial incentive applies to systems larger than 200kW for all locations in NY State other than Long Island, as well as not-for-profit organizations and government buildings with systems of up to 200kW.

There was also the Federal Investment Tax Credit (ITC), where home and business owners purchasing a new solar power system were eligible for a federal tax credit amounting to 30% of the cost of the system. However, it expired starting 2017, and there is no news of whether Congress plans to extend the tax credit.

Last year, solar became the leading new energy source in the United States, with new installations doubling to 14,600 megawatts of capacity. It made up almost 40% of new power arriving to a grid, surpassing coal, wind and natural gas. Solar farms made the biggest most significant contribution, generating over 1 megawatt in large-scale production. In all, 22 states each added over 100 megawatts of capacity. California continues to be the hub of solar and other forms of renewable energy.

NET METERING AND
RENEWABLE ENERGY CERTIFICATES

Renewable Energy Certificates (RECs) are credits that a producer of renewable energy gets that document their involvement in the creation of renewable energy through either wind or solar energy. Other ways you may see them described are Tradable Renewable Certificates, Solar Power credits, or Green Tags.

The certificate's currency is a megawatt hour. (1 megawatt-hour = 1,000 "kilowatt-hours," which is the term you are most likely to see on your power bill.)

There is a similar certificate called an SREC that is specific to power created by solar panels. With mandates in certain states for generating renewable energy, these credits have become the currency that bails utilities out of hot water. The energy that you create with your solar panels may end up being purchased by your utility to prove to state authorities that they are going green enough to offset the fossil fuel-based energy they create and sell to everyone else.

I don't mean that the utility does not want to go green. The utility may be in an area where there are no good options at present for creating wind or solar power. And, to meet the mandate, the utility must purchase RECs from some entity that can produce renewable energy. This example underscores for you that there is more than one way to skin this solar cat. You can supply your own energy with solar, but you can also provide renewable energy certificates for organizations that need them for other reasons like meeting mandates.

How Much Do Renewable Certificates Cost and Where Are They Traded?

RECs are traded at SCRECTrade, and the price range is $180-$260 per certificate. Solar is the only certificate market they trade in. Although they are not the only market for solar certificates, SCRECTrade may be the largest. They are based in San Francisco and started in 2007.

Net Metering

Net metering is defined a little differently in each location where it has been used. In general, it means the ability to account for electricity exports to the grid from individual energy producers. Thirty-eight out of fifty states, including four territories, have adopted net metering policies that allow renewable certificates to be created and state Renewable Portfolio Standards to be met. Seven states have not adopted net metering policies—Arizona, Georgia, Hawaii, Indiana, Nevada, Maine, and Mississippi.

Although there are a few exceptions, most places that have net metering allow for the credit to the energy creator to occur at full retail compensation. That means if you have a solar collector system and have net metering, the amount of energy you exported to the grid from solar during the following month will be credited to your account at the full retail price sold by your utility. Not a bad incentive and a great way to offset rising energy costs with renewable energy.

Now, while solar may be the most common way to take part in net metering, it is not the only way. Other renewable systems that qualify—such as wind, biomass, natural gas micro-turbines, and methane digesters—can be used to offset the energy bill you receive from a fossil fuel guzzling utility.

Who Buys and Sells Renewable Certificates?

The participants in this market include anyone needing to supply proof of being a party to a renewable energy transaction either for marketing or regulatory purposes including owners of the renewable resources, electricity suppliers/utilities, projective developers, and renewable installers.

The Renewable Certificates Program Is Here to Stay

As we discussed earlier, products that do not currently provide a robust economic value but would be good for society and the world, need to be initially supported through a program of government sponsorship or credit. Eventually, these products and programs will take root. But, we cannot afford to wait for free market forces to take over and drive prices down as we enjoy the societal benefits.

Initially, with any new endeavor like renewables, there will be few competitors making products and providing services. That is because there isn't enough of a market base for many people to get involved and make much money at that stage. It's a chicken and egg scenario—you need the market to be profitable to incentivize involvement and provide competition. However, the market will not become even marginally profitable unless some people start buying the product and using it, then recommending it to their friends. This is a delicate balance in a product or service's early life cycle, and it can quickly get derailed if there is not a support system to artificially make the product profitable as early as possible.

The government's rationale for taxpayers to underwrite renewable technology is that it is ultimately in their best interest to have a robust renewables sector. Looking out for the long-term best interest of the entire group is not easy—winners and losers must be selected by the government.

Currently, the U.S. government is heavily subsidizing the renewables industry in hopes that prices will come down significantly and then competition will take over and drive prices down yet further. The subsidy is coming in various forms, from credits and rebates to the creation of the renewable certificate program that we have been discussing.

The underwriting of renewable technology will continue into the foreseeable future because the industry continues to need bolstering even though competition had increased somewhat since the early days when only tree huggers were engaging in this debate. Long term, this support may not be required, but it is likely that the certificate program will continue since it now has companies and members who have made a profession out of trading certificates.

HOW TO COMPARE YOUR SOLAR STORAGE OPTIONS

The most challenging factor when it comes to adopting solar is getting started. With so much information out there about the benefits of solar, it may not surprise you to learn that there are fly-by-night operations that are trying to take advantage of the buzz.

So, where to start? It does not seem right to run with the first company that happens to cold call you. Solar is a lot like lighting these days. Many vendors are selling products that all seem to look the same. It is difficult to know who is providing a superior product at the best price versus who is selling junk. In many ways, we are all starting from scratch.

Is There a Solar Products Quality Counsel?
Will DSIRE Recommend Vendors?

Many state agencies have an office dedicated to dealing with solar. In states where solar is being supported at the state level, you can ask that state office for guidance about approved manufacturers and installers. However, if you happen to be in a state like Georgia that does not promote solar, you are pretty much on your own. With the technology developing so rapidly, you need to be able to evaluate the state-of-the-art technology. It could be that the cheapest product is not the best product for you. Or, it could be that the most expensive product is not the best for you. There is no substitute for exploring the numbers and understanding how the different offerings by vendors will apply to you. It is also important to consider and research the track record and market longevity of specific vendors.

There is probably no avoiding some detail work on your part to evaluate the systems that are out there.

Batteries are to solar power as the right hand is to your left hand. While you can get away without having battery back-up, it would limit the utility of the system so much so that it is pointless to consider pricing a system without also pricing a storage or battery system.

To correctly size and price the battery component of your solar power system, there are some terms with which you need to become familiar:

Capacity and Power Rating

This is the total amount of energy your battery can store and hold for you. It is measured in kWhs. Every battery has such a rating that describes how it behaves over time. How stackable the system means how many kWh-rated batteries can you attach in series to achieve your backup goals. The power rating is measured in kW and is the instantaneous power available through the batteries discharge.

DoD (Depth of Discharge)

You also need to know the depth of discharge because that describes how quickly the battery will deplete when called upon to supply power. Example: If the manufacturer claims a five kWh battery has a 90% DoD, then you will get 4.5 kWh out of the battery before it needs recharging.

Round-trip Efficiency

The higher the round-trip efficiency of your battery, the better the value. If you store ten kWhs in the battery and it yields 90% round-trip efficiency, you will be able to pull nine kWhs out of the battery when you need it.

Warranty

Warranties, of course, differ widely between manufacturers. As you consider different options for batteries and solar power systems, carefully weigh the warranties offered by the various manufacturers side-by-side. While new technology companies may provide cutting-edge products, it may be prudent to go with an established brand because of their longevity in the marketplace and therefore their ability to honor their warranty over the long haul.

In summary, the battery system you choose will have an enormous impact on how much savings you will achieve through coupling solar with demand response. For instance, if you have enough battery backup to power your equipment for three to four hours (the length of the most common demand response curtailment), it will be very easy for you to participate in such a program and get the benefits.

However, you will be already dedicating a lot of real estate to your solar panels, whether it is on your roof or otherwise. The batteries will also take up a good bit of space, so you need to be wary of that and look at options for stacking batteries vertically.

Battery Life

For most uses of home energy storage, your battery will "cycle" (charge and drain) daily. The battery's ability to hold a charge will gradually decrease the more you use it. In this way, solar batteries are similar to the battery in your cell phone—you charge your phone each night to use it during the day, and as your phone gets older, you'll start to notice that the battery isn't holding as much of a charge as it did when it was new.

Your solar battery will have a warranty that guarantees a

certain number of cycles and years of useful life. Because battery performance naturally degrades over time, most manufacturers will guarantee that the battery keeps a certain amount of its capacity over the course of the warranty period. Therefore, the simple answer to the question "how long will my solar battery last?" is that it depends on the brand of battery you buy and how much capacity it will lose over time.

For example, a battery might be warrantied for 5,000 cycles or ten years at 70 percent of its original capacity. This means by the end of the warranty; the battery will have lost no more than 30 percent of its original ability to store energy.

Manufacturer

Many different types of organizations are developing and manufacturing solar battery products, from automotive companies to tech startups. A major automotive company entering the energy storage market likely has a long history of product manufacturing, but may not provide the most revolutionary technology. By contrast, a tech startup might have a brand-new high-performing technology, but less of a track record to prove the battery's long-term functionality.

Whether you choose a battery manufactured by a cutting-edge startup or a manufacturer with a long history depends on your priorities. Evaluating the warranties associated with each product can give you additional guidance as you make your decisions about solar.

CHAPTER SUMMARY

Solar energy costs have fallen significantly, and their efficiencies in energy production have risen in recent years. By investing in solar energy production, you can offset increasingly expensive energy purchases; you are making now from your utility. Batteries and net metering are vital components that expand solar's versatility and cost savings potential.

QUESTIONS TO THINK ABOUT

1. What are the total power requirements for my facility?

2. How much roof space does the facility have?

3. What is the efficiency of the solar units being considered?

4. How much battery power will I need?

5. Have I found 3 suppliers to contact about analyzing my options?

Chapter 8

Energy Management Systems

Another powerful tool in your arsenal to cut costs is having an energy management system (EMS) in place. An EMS done correctly will serve as a constant reminder of usage and a rallying point for change. If you later focus on consumption reduction, an EMS can serve as a validator that your cost reduction efforts are paying off.

The continuum of costs and the complexity of these systems vary greatly. I am advocating only the low-tech, simple to install end of this spectrum. It is very easy to become inundated with useless information. Such systems end up like the Strato-stepper 1000 you bought on TV that now serves as your staging area for the clothes closet.

COMPONENTS OF AN
ENERGY MANAGEMENT SYSTEM (EMS)

First of all, an EMS does not "manage" anything. It reports, and then it is up to you to manage. In its purest form, an EMS "sub-meters" your operation and reports the amount of energy each area of your operation is using, both in real-time and historically. By comparing usage patterns for one block of time over successive days and weeks, you can get a sense of where to focus your attention to see the most improvement.

In that sense, an energy management system serves to prioritize the questions you need to ask. And it gives you the data to ask those questions with authority. The data will provide you with an objective reason for asking someone in operations why the HVAC system usage went up during the last three shifts, for example.

Then, you can create a small team to drill down, understand the problem further, and recommend solutions. Once a chosen solution is in place, you can use the EMS to verify and validate the decision. Having the watchful eye of an EMS on site means you are providing a framework for accountability going forward.

Hawthorne Effect

Accountability comes from knowing others are paying attention to our details. It's easy to fool ourselves with rationales and half-truths to avoid action.

In business, we are usually operating without a metronome, and no one's really watching what we do on a day-to-day basis. Out of sight means out of mind. If we think no one is watching, we all tend to ease up and not pay attention to details. The gyms are filled with people living up to their highest and best practice standards on January 1. But by February 1, gyms are empty again. Why? The Hawthorne Effect.

The Hawthorne Effect says that our performance improves when we are being watched. Elton Mayo observed this in FC Hawthorne Works, a Western Electric factory outside Chicago. The Hawthorne Works commissioned a study at the facility to see if lighting changes influenced productivity. They discovered it didn't matter whether the lighting went up or down, the productivity still improved!

The Hawthorne Effect means people work harder and are more accurate when they believe their actions are being observed and noted. When we understand that the actions we are taking are being recorded or criticized by others, we pay more attention and productivity naturally improves.

This can be a powerful tool in energy management. With the stresses and volume of work being done in a modern manufacturing plant, no one will notice if you don't follow through on energy projects without plainly stated oversight. Since the product is getting out the door, and the energy bills are getting paid, no one in management is going to ask about the energy project.

In our case, that means we should make the key players aware their data are being tracked.

How do we use the Hawthorne Effect to make the most out of an EMS that is already in place? The simplest way is to track the sub-metered data over time and broadcast results graphically to those in charge of reducing energy costs. Also ask the floor manager in each area to put together a two- to three-person team to study the data and offer solutions for consumption production to be reported to management.

Where is the Low-hanging Fruit?

Deal with the 80/20 of energy cost reduction. The devices that consume the most power in facilities are also the places to find the greatest amounts of wasted energy.

The extreme energy consumers in most plants, other than production equipment, are HVAC, lights, and compressors. Not only are these areas subject to very little control and management, but they are also areas where the technology is changing rapidly and thereby offer great options for improvement.

The fact that these areas are not controlled much means they are often sinkholes for waste. Lighting costs for example: without automation, lights are turned on and left on. Heating and cooling also gets adjusted for personal taste, and not for optimal energy savings. Air compressors are usually either all on or all off regardless of whether they are needed at the moment for actual work.

Simple vs. Complex

Although Moses didn't carry this law down from the mountain, I believe there is a right way and a wrong way to use energy management systems. The most straightforward system is usually the best system, and it begins with isolating waste. A more complex system that tries to track everything is doomed to be crushed by the sheer weight of its data.

The problem with the more complex systems isn't that capturing energy minutia is worthless; it is just that the returns are too small for the effort you will expend. It is too much to expect a plant, thinly staffed already, to make use of all that new energy data.

Before you invest in energy management software and hardware, you need to have a clear idea of where monitoring the data will likely lead you. For example, with lighting, you are monitoring your lights to provide a baseline for gauging future technology improvements. Also you are monitoring to get people to pay attention and turn off lights when they are not in use.

If you can't simulate the impact on the operation of measuring the energy usage on a particular piece of equipment, then do not invest in monitoring hardware. All monitoring devices need to earn their keep. Do not let yourself get pulled in by the salesman and buy more than you need or can effectively stay on top of and use. It is unlikely a system like that will ever pay for itself.

The costs of all-encompassing systems can be staggering, and the extra money you will pay for going down to a micro-level with your monitoring will get lost in the overall installation costs. It is unlikely there will ever be an audit of the effectiveness of the system you have installed, and you will have created a process that you will feel you need to monitor—passing the data down the line but mystifying a staff already overburdened with too much variable data.

There is a high unaccounted for cost associated with collecting data. Not only does it cost money to harvest the information, but someone must slice and dice it, report it, and then follow up on it. If the data is not purpose driven, it just becomes background noise.

So, I am urging you to keep things ultra-simple with your EMS. Start slowly and measure only the large energy users in the plant. Resist the urge to put data readers on everything that moves, as the vendors often recommend.

By focusing on the large energy users only, you can prioritize and affect some quick successes by removing waste. Momentum is a critical element in all projects and making your first foray into controlling your energy a successful one.

Keep in mind that no system like this works unless it is vigorously pursued. You can't install an EMS and then forget about it. You must make the EMS a priority, and you must follow up with it regularly until the dissemination and action on the data become routine.

Automating the reporting is critical. Only choose systems that offer template reports because coming up with your own can be very time-consuming. The most powerful programs are in app format for multiple users with a focus on smartphones.

How to Get the Maximum out of
An Energy Management System

Ultimately, an energy management system is only a vehicle for you, a vehicle that helps you keep your eyes on what matters most.

Our goal with any expenditure in business is to get the elusive return on investment. There is probably always some measure of return on investment with any expenditure, but it is only the returns that bring the highest dollar-value in the fastest amount of time that we pursue. With an energy management system, your returns can be high, but mainly over the long term. Why is committing to the long-term nature of the investment important?

"Energy management" is more of a commitment to our belief in the power of consistency. Energy management works the way it was intended when the data are reviewed daily. The salesman will probably sell you the EMS equipment based on how easy it is to install and maintain. But, therein lies the problem; to reduce your energy costs, you must commit your focus on reduction a little every day come hell or high water. We want easy installation, and we want the EMS data to bother us until we take action.

The EMS will be most valuable to you if you look at the data it produces like a mental metronome or a rallying point for process improvement.

Unwatched systems, whether we are talking about plant energy expenditures or government bureaucracies, become wasteful in a hurry. That is not a comment on the laziness of the people watching it; it is the nature of systems of any kind. In other words, we have choices of where to place our focus, and if we don't make energy data front and center, then it will be an afterthought and treated as such by the people in your organization.

The power of focus is simple to grasp. If you weigh 250 pounds and want to weigh 200 pounds, it is very easy for that desire to get lost among the myriad other things going on in your life. If you have no current health problems that are screaming at you to change your weight you probably won't lose the weight. Why? Because there is no feedback system in place to remind us to take those daily actions.

However, if you bring in a scale and make weighing yourself unavoidable—like putting it right at the door, you usually go in and out of—you will force yourself to be reminded of your current state (250 pounds) and by association, the state in which you aspire (200 pounds). If you then make it easy to mark down on a chart what you weigh every day, you will be giving yourself a picture of your own consistency and a chart of your progress. Taken a step further, if you make the chart visible to others by posting it on a wall, you have added yet another dimension—peer pressure, or our desire to impress others with our consistency.

So, how does this relate to saving money on energy? Your energy waste at your factory and in your buildings correlates well with the 250-pound person who desires to weigh 200 pounds—no matter how efficient you think you are, you can make your energy consumption dramatically more efficient when you pay attention to energy efficiency.

If you don't currently have an energy management system,

you are guaranteed to be inefficient. People usually believe their energy spending "is what it is." They think the equipment and the lights use whatever they are rated to use and that is all there is to it.

Unfortunately, that is not the case. With great power comes great responsibility. Set up monitoring systems that you review daily.

If EMSs are the Big Mac, then daily management is the special sauce that makes the Big Mac. Having a system that you don't use and monitor every day gives you the bragging rights. It will only impress people who do not currently have such a system. To make it an organic tool for improving things, you must pull the data and review it with others every single day. You must post the data and the trend analysis frequently. You must post it at a place where people cannot avoid seeing it. Making the data stand out is the only way to get everyone in on the act and drive results. You will be amazed at how this simple system yields results over time. This is the basic blocking and tackling of goal attainment that you already know.

Once our beliefs are dashed about the fixed nature of the "way things are," we are more open to improving in a lot of other areas too.

SOURCES OF FUNDING FOR ENERGY PROJECTS

One of the major stumbling blocks I see in companies is their inability to actively manage their energy costs. After you have done everything you can to reduce costs without spending money—that is the action steps we have talked about in this book—then it is time to go after energy efficiency. This is going to be achieved by cutting out waste. Every system that I have ever seen has inherent waste.

It is because of my abiding faith in the idea that all systems have waste that I say with 100% certainty, a small investment in a monitoring system will pay huge dividends. Out of sight, out of mind. If you are not looking at your data, there is no way you can improve. If you are not looking at your data on a regular basis, it will not be front and center in your mind, and there is no way you can control it. The data must be bugging you. Set up systems where you cannot avoid seeing it.

To initiate an efficient monitoring system, you should have a mirror image reflection of your utility meter close by—either on your phone or desk. The main reason we waste money in energy is because we do not have the data staring us in the face. There will be no way to generate a question without having existing data to back it up. Asking, "can we reduce our lighting expense?" Is not nearly as powerful as "I noticed we are spending money on lights during a time when we are not working." Believable, close to real-time consumption data give us that.

This is all about generating questions. Discrepancies in the data will generate those questions. If we measure what the standard is and then contrast that with what is happening, we will generate questions. Then it is our job as managers to share those questions with the people who can answer them. Employees do not inherently want to waste, but if they are not being rewarded for paying attention to the details, they will not be as concerned with those details.

It is never a good idea to introduce questions to employees in a threatening way. The people on the floor doing the job are not trying purposely to be wasteful. They are not going out of their way to spend more money than is necessary. However, we are all at the mercy of the people who are watching what we do. If we do not feel like people are watching our work, we tend to not pay as much attention to the details of the work. We require a feedback loop in order to do our best with anything. As soon as people on the floor are aware that management is looking at real-time data, that fact alone will reduce consumption.

So, it is incumbent upon us to collect the data that we need and share the data with employees. We cannot expect them to self-correct and do this without feedback.

The feedback loop and the backing up of the questions with data are the two keys to reducing energy costs in the short term and long term. It is in our best interest as an organization to cheaply and efficiently get this data where it can be reviewed and acted on as soon as possible. There are programs available whereas a business you can get financing either directly from a lender or through your utility to pay for such a system. That is the subject of the next part of this book.

What Kinds of Projects Merit Independent Funding?

We are talking of course about energy management data now, but any project that leads to greater efficiency can be considered as part of this scenario. That means if you have equipment that needs to be replaced that will result in lower energy costs, there are lenders as well as your utility that have an interest in helping you achieve that goal.

It is up to you, however, to put the project together and sell it to whichever entity is likely to finance it, whether it be the utility or third party. The advantage of going with the utility and on-bill financing is that you do not have to engage another entity to get the job done. Also, your utility understands energy and can help you with the economics of your particular project.

The best way to begin is to reach out to your utility and find out if they would be open to sponsoring an energy management project. Contact your rep, or if you do not have one assigned to you, reach out to the business development office at your utility.

CHAPTER SUMMARY

A simple energy management system (EMS) can save you significant funds by uncovering waste in your operation. Waste is exposed by monitoring when energy is used and comparing that with expectations. An EMS can also serve as a framework for understanding the value of future efficiency improvements.

QUESTIONS TO THINK ABOUT

1. Is there an energy management system in place now?

2. What is the unit cost of demand from my power company?

3. Do I have electric heat load or air handling I can manage?

4. Have I located at least 3 alternative energy management systems to price?

5. Who at my company will manage the program and review data?

Part III

Free Money

Chapter 9

Sales Tax

SALES TAX EXEMPTIONS

The most straightforward actions can sometimes yield the biggest results for the least investment.

Getting a sales tax exemption has nothing to do with being a charity. States exempt all types of businesses that are not charities from a variety of things—incentivizing manufacturing business, jobs, and revenue into the state to be taxed in other ways.

Most of you reading this book will be in the manufacturing arena. Depending on your state, this opportunity may or not apply to you. There are, however, some definitional nuances we will talk about to check out if an exemption is a fit, because it is not always straightforward and it does not happen automatically.

What Does Being Sales Tax Exempt Mean?

Being sales tax exempt means, you do not have to pay sales taxes on some or all of your utility consumption. It is based on some set of state-specific criteria. Most people in business have no idea that their utilities can be made sales tax exempt.

Why? Because the state departments of revenue do not broadcast it.

As the default position, all utilities tax the sale of their product. They collect the money for the state, city, and county at their mandated percentages and send that money to the respective governmental entity. The utility makes no money on collecting the tax. So, they are indifferent to whether you claim the exemption or not.

You must "claim" your sales tax exemption and sometimes even insist on it. You must do it yourself because it will not come to you and you won't be reminded to claim it. It will be no surprise to find that your state will gladly keep all the money you send them.

In our litigious society, some utilities have taken the position that it is a bad idea to even discuss the possibility with you that you *might* be sales tax exempt. Thus, sales tax will stay on your bill until you prove to either the state or to the utility that you deserve to be exempt.

WHICH BUSINESSES ARE EXEMPT AND WHY?

Only manufacturers and processors in certain states are exempt. It all depends on the state and their approach to attracting business. All states look at other states, especially neighboring ones, as their competition in that way. The reasoning the states use is the belief that collecting fewer sales tax on utilities can lead to a broader base they can tax in other ways.

Since sales tax can be 7-9% of your utility cost, it can be a very large number and make the difference between profit and loss in some large energy consumers.

If You Qualify,
It All Depends on the Definitions

The definition of your business and the specific mission of your business is very important in determining your sales tax exemption eligibility. Manufacturing is defined in most cases as taking component materials, adding value to them, and then re-selling them. This is a very broad definition that includes many more companies than you would think. Processors often feel they are not in this exempt class, and in many cases, they are wrong.

It is not important whether you call what your business does "manufacturing" or "processing," all that matters is fitting the state's criteria. So, pay strict attention to your state's technical definition of which businesses qualify.

As in all areas of law, you can find cases that support making a claim for exemption status and cases that support not making a claim. It is important to be honest and clear in your definitions and if you have questions, ask your state department of revenue. You have a responsibility to seek out all forms of savings and make them work.

Once you have achieved exempt status and the tax is no longer on your invoices, it is time to reach back in time and claim a refund of taxes you have previously paid.

REFUNDS FOR OVERPAYMENT

As a nice bonus, most states that offer an exemption from sales taxes also allow you to file for a refund of overpaid taxes from the past; three to four years in most cases. Once you have filed for and achieved sales tax exempt status, you can then apply for the refund. This part of the sales tax trail can be tedious and time-consuming. Many states are customer service oriented and make the process quick and efficient. Other states force you to make a Faustian Bargain to get things done.

Some utilities will simply credit the refund amount on your bill, but most require a separate filing with the state to get your money back. When the state requires a filing, they will require you to prove to them you paid the tax. This usually means sending in copies of actual bills. Attaching a summary spreadsheet and a claim for refund is required. The processing can take months if not years to process.

The length of the process and hassle factor dissuades some less intrepid companies from even filing for a refund. However, the amount of these refunds can be substantial. Do not become frustrated with what seems like government stonewalling—it is usually a matter of limited staffing. Giving back money is not an area the government adds more staff to handle.

STATES WHICH ALLOW MANUFACTURERS TO EXEMPT SALES TAX ON UTILITY BILLS

Manufacturers in 37 states are exempt from paying sales tax on utilities. Some states require a form, and some forms are a hassle. Some states mandate the exemption for certain Standard Industrial Classification (SIC) codes, and there is no additional paperwork required. Each state that offers an exemption for certain SIC codes is unique and must be researched in detail before filing.

You would think there would be a collusion of minds, and there would emerge a "best practices" among states, but there is not, the states all have their own way of doing things.

This list changes frequently. As of 2018, these are the states that exempt manufacturers from sales taxes in the U.S. in some way shape or form:

Alabama	Iowa	Ohio
Alaska	Kansas	Oklahoma
Arkansas	Kentucky	Pennsylvania
Arizona	Maryland	Rhode Island
California	Massachusetts	South Carolina
Connecticut	Michigan	Tennessee
Colorado	Minnesota	Texas
Delaware	Missouri	Utah
Florida	Nebraska	Wisconsin
Georgia	New Jersey	Wyoming
Idaho	New York	
Indiana	North Carolina	

States that exempt manufacturers from sales tax (2018)

During the last U.S. recession in 2008, some states with exemptions in place took them back and started taxing manufacturers again. The state legislatures panicked and forgot the original reason for the exemptions in the first place—to incentivize manufacturers to locate in their state, thus providing jobs and more taxable income and revenue. So, check the current exempt status in your state.

Some states, like Wisconsin, have an exemption form to file. You receive a certificate back, and then you send that certificate to your utility. Other states like Texas require you to write a letter requesting the exemption with no clear format required. There is no one size fits all—you must check with the state department of revenue at the time you are filing for the exemption to find out about their current procedure.

The process can range anywhere from simple and straightforward, to requiring complex modeling of the energy usage. There is no one in this loop insisting that you do any of this.

You need to act quickly once you file for exempt status. Some states set a timer on the refund request process. If you do not file for the refund within a certain number of months after you file for exempt status, then you may not be allowed to collect your refund at all in some cases.

What is The General Procedure for Becoming Sales Tax Exempt?

Sales tax exemption status is a several-step process:

1. Determine if your company qualifies for the exemption—the best way to find out is to search for your status form on Google. Going to your Department of Revenue's website to find them will be an exercise in futility. Try the search phrase, "Manufacturers sales tax exemption utility form (your state or province)."

2. Gather the information requested by the form and fill the form out. Send it in along with any required back-up documents or copies of documents to the address listed on the form. Although some states have electronic filing, filing electronically can often send you into a black hole. I recommend mailing the form in with a return receipt so you can make sure they received it.

In my experience, all of this is easier said than done. Usually, when you send in the forms, they do not automatically get processed. Reach out to the entity in question early in the process, whether it be the utility or the government department of revenue, and make sure the forms you sent have been received and are in process.

I have seen forms packages that languish for months and years because a form didn't have every "T/I" crossed and dotted, while the package was languishing on a bureaucrat's desk. Do not have faith in the process. They are not looking out for you or your business. It doesn't take much wrong with it to get a bureaucrat to delay an application. This is a system, and these are people not in a hurry and feel no special obligation to let you know where you stand.

You will have to monitor progress yourself.

What is The Procedure for Applying for a Refund?
The great news is you are usually entitled to a refund. The bad news is you have more paperwork to fill out.

The procedure for filing for a refund of previously paid taxes is usually a lot more time-consuming. Some states and some utilities make it super-easy by totaling and crediting the refund on your next utility bill. Some, however, make the process as laborious and litigious as possible, to discourage you from asking for a refund.

This is How to File for a Sales Tax Refund
1.   File the necessary forms to process the refund of previously paid taxes. Google search phrase, "Manufacturers sales tax exemption refund form utilities (your state or province)."

2.   Gather the information the form requires and fill it out. Send the form and any required back up documents to wherever is listed on the form.

Again, this is a process that can get bogged down in red tape and move slowly. No one on the other side (utility or department of revenue) cares that much whether you get your refund or not, so things can move to a snail's crawl and stop if you do not stay on top of it.

The philosophy that wins is "polite persistence." You can always ask a question frequently if you do it politely and with respect. Be sure to get the name of who you spoke with and ask them when they expect things to be done. Follow up accordingly.

It makes the most sense to be aggressive when it comes to claiming what is yours in life. As the great hockey player Wayne Gretzky says, "You miss 100% of the shots you don't take." In most states that offer the exemption, the definitions of what constitutes a manufacturer or processor are very broad. If you can defend your position, it makes sense to make a claim.

Sometimes Companies Will Not Be Completely Sales Tax Exempt

I have seen companies that are savvy and sign up as a manufacturer to be exempt, but then guesstimate too low a percentage of exemption. You need to be accurate, not conservative.

This is where a predominant use study comes into play. Engineering or maintenance departments are often given the task initially of estimating the percentage that should be exempt. Often, they opt for a very low percentage just to be "conservative."

I am always suspicious of this when I see percentages given in round numbers like "50%" or "75%." That is rarely accurate.

And There's One More Thing—"Utility Holdback"

Many states allow "collecting entities," like utilities, to get paid a little for their trouble. It's not as much as states charge for their sales tax, but it is refundable money none the less. Focus on getting your refund from the state first, because the holdback will be more difficult to attain. It may be difficult to even find anyone at the utility who can talk with you about this.

Again, the onus is on you, the intrepid energy cost reducer, to do your research and then be persistent. You may get a range of responses from "what are you talking about?" to "we don't do that." Doesn't matter. Pursue it anyway because it is there. This is money you are entitled to.

You will find that everything to do with sales tax refund is based on what you ask for. If you don't know to ask, utilities feel no obligation to help you with it.

CHAPTER SUMMARY

Many states will give you a break on sales taxes paid on electricity and other utilities. A refund of previously paid taxes is also available for three or four years before your tax-exempt sales status. It can be a hassle to file, but it can be a nice free boost to your bottom line.

QUESTIONS TO THINK ABOUT

1. Does my state allow sales tax to be exempted for my type of business?

2. Which utilities are exempted? (electricity, gas, water/sewer)

3. What is the procedure for claiming the exemption and refund?

4. Is a predominant use study of my facility required?

5. What forms must be filed with the utility or the state Department of Revenue to achieve the exemption and refund?

Chapter 10

Billing Line Items You Can Remove

RATCHETED DEMAND CHARGES

Here's one place you could be sitting on a pot of gold and not realize it. If you have equipment that you have taken out of your process or if you have lighting that you have upgraded, you need to let your utility know about it.

On many utility rates, one of the critical elements is the demand ratchet. On a ratchet, you set a peak demand sometime during the year, and then you live with that demand level for the next 11 months. That means the utility will continue to charge you at the peak demand rate even though you may be consuming only half of that amount.

Ratcheted rates are the norm now and were the only thing utilities had to offer for a long time. The reason utilities like to bill this way is simple—billing you for the maximum demand you create in each month and the highest demand you register in each year, is a planning tool for them.

The utility gauges what you are likely to use in the next calendar year by the peak demand you set. What you used once you could conceivably use again. For a utility, planning is not just a good idea, it is required by state public service commissions charged with managing generation load growth.

So, utilities see your maximum billing as a barometer and therefore a harbinger of things to come. By billing you through this ratcheted system, you are paying ahead for the generating capacity the utility believes you are telling them you need.

However, this cuts two ways. If you are downsizing and you prove to the utility you have taken equipment or have otherwise reduced your demand on their system permanently, they may be able to reduce your demand cost. Your demand cost can easily be over half your bill every month, so it pays you handsomely to investigate what efficiencies have gone in place as equipment is replaced. Your utility has no way of knowing that you did that unless you make them aware of it.

Document the decreases in demand thoroughly, including taking pictures of any machines, etc. that have been taken off line and that you do not plan to reactivate. That information needs to be sent to them in the form of a letter, documenting these facts along with a request for them to reduce the amount of demand they are charging you now.

I have seen bills reduced significantly by using this strategy. You need to make sure you understand the way the current rate you are on works first. If the rate has this ratchet formula built in, this strategy will reduce your monthly bill in the next thirty days. If there is no ratchet—like in the case of most time of use rates—your demand will reset each month automatically, and there is no advantage to pointing out the decrease to the utility since it reboots every month anyway.

Paying attention to changes in lighting is an especially powerful way to use this process to reduce your bill. With a lighting project, not only are you reducing costs on the go-for-ward but you also can get the old equipment factored out of the billing equation. This is seldom figured into the project economics, but it should be. And, not many utility reps understand this principle so you will most likely have to educate your utility that it should be your right not to have to pay for the demand that you are no longer creating on their system.

SYSTEMS RENTALS/FACILITIES COSTS

Just because an item is on your bill does not mean it belongs there. Saving big money is very much about having the confidence to ask questions and not being satisfied with generalizations and utility assurances that you have the lowest cost in their service territory.

Let me tell you a little story that will explain what I am talking about.

When I first started out in manufacturing energy management in the 1990s, my first project was to look at our company's electricity charges and see if all of it "still made sense." I had no idea what I was doing, but I stumbled upon a charge on one of the bills that I didn't understand, and our utility rep didn't understand either.

That line item was a "facilities charge" of over $3,000 a month. It had occurred every month for years, and no one had brought it up or questioned it. With a legitimate-sounding name like "facilities charge," who could blame anyone for not checking?

As it turned out that charge was for what the utility deemed "extra" facilities—that is, charges for transformers and power delivery equipment specifically related to our plant; those required over and above the monetary value of the facilities that had been baked into the rates we were paying.

Someone involved in energy at the plant and someone at this utility had created this charge years ago, but it had no expiration date set, and no one ever gave it a second look. The circumstances that created the charge were long since forgotten.

The bottom line was the charge did not belong on the bill anymore. We worked together and removed it, and the utility rebilled the account and issued a refund for the incorrect charges. (When a utility "rebills" an account, it is a way of reconciling a billing mistake. It means they run the billing again in through their system to see what the total charges would have been without the error. Then, any difference in the simulated charge and the actual bill can be issued as a credit or a debit on the next invoice.)

Do not be afraid to challenge line items on your bill that do not make sense. If they do not make sense to you, they may not make sense to your utility rep either.

UTILITY-SPONSORED GREEN MEASURES

Although being greener is a good thing, in some cases of utility-sponsored green measures, it might not be good for you.

Many states are mandating that all categories of customers, including manufacturers, help pitch in and pay for incentives to add facility-based renewable energy. This means adding solar, wind power, and reducing the amount of energy you consume. While these are well-intentioned programs, they do not apply to all ratepayers, and you shouldn't have to pay for them if you are not going to use them.

Often these charges show up as a line item on your invoice under a nice name like "energy incentives." If your company cannot or will not take part in these programs, you do not necessarily have to put your money in the state kitty. Ask your utility or your state renewable energy office if these are charges you can opt out of. If so, you may have to sign a document saying you won't apply for the program benefits. Each state's program and what demand reduction options it sponsors is different.

Once you have done this, you can get the line item taken off your bill. You may even be able to get the money you've already paid returned to you.

How to Question the Line Items on Your Bill

If you are like everyone else, you do not challenge the details on your utility bills. Utility bills are often coded in impossible to understand terminology.

We are, of course, encouraged to learn what the terms mean by our utilities. On their websites, many utilities have "bill explanations." Because of their past as all quasi-government entities, we tend to accept whatever the utilities tell us as just the way it is.

When items on a bill are not challenged, nothing ever changes. These extra charges, many of which make no sense, get added to the costs of running our businesses with no pushback.

Names for the charges are chosen not to clarify and explain, but to keep down the customer service calls questioning them. If we don't understand, we don't ask because we don't want to feel dumb because we don't know.

But what is being done with the money for "Storm Recovery?" We all tend to play along and don't ask these questions, assuming someone else wouldn't have put the charge on the bill or would not have been audacious enough to charge a large group of people for something if there were not a legitimate reason to do so.

If we challenge these little charges, sometimes we walk away with a win.

You must ask and be persistent to fight against the indifference from the customer service person you are likely to talk with, but sometimes you can opt-out of these charges.

There is Never a Downside to Asking Questions

If you are polite about it, you can ask most any question and get away with it. Given that, it makes zero sense to suffer in silence—just ask. However, do not be automatically satisfied with a general answer to a specific question.

The litmus test of a great question should be: If this question were answered accurately, would it give me all the information I seek? If it is possible to answer the question in a general way, most people will answer it that way.

Take the time to pose your questions to your utilities after some deliberation. Be prepared that your utility rep or the person who you get on the phone may not know the answer no matter how explicitly you phrase it. Prepare also that it is very likely you will have to insist that they follow up and get the question answered by others. They may even be offended that you are questioning things. Press on anyway.

Can You Always Get These Seemingly Random Charges Taken off Your Bill?

It all depends on the utility and the line item. There are no givens. However, if you do not ask, you will not get. You must have the attitude that utilities make mistakes. Just because one of your charges sounds very official does not mean that it necessarily applies to you.

One of my objectives with this book is to get you thinking as I do—nothing in the utility world is cut and dried like the powers that be would have you believe. You must question everything relentlessly to succeed.

It is usually easier to just breeze by these unintelligible charges on our bills. Realize though that a human being or a group of them put that charge on your bill. When we humans are involved, mistakes are inevitable. So, with respect for your fellow humans and their fallibility, question the charges and do not be satisfied with easy answers.

When you ask questions boldly, you always learn something.

CHAPTER SUMMARY

Charges appear on your bill as line items that look very official and permanent. But you can often opt out of them by making a request.

QUESTIONS TO THINK ABOUT

1.   Do I understand why I am being charged for each line item on my bill?

2.   Have I asked my utility rep if I can opt out of any of the charges?

3.   What is the opt-out procedure?

4.   Is there any reason I would not want to opt out?

5.   Can I get a refund?

Chapter 11

Combining Meters

Are you now receiving more than one bill from your utility provider every month for different meters? If meters are located on the same piece of property, you may be able to have the utility combine them or consider them as one meter for billing purposes.

The problem is each of your meters is at a separate rate.

WHY SO MANY METERS?

First, each meter carries with it a "customer charge" of $30 up to several hundred dollars a month—that is, every month, forever.

Your utility is often running your usage through the early and more expensive steps of these rates for each meter. You are being billed for energy at different prices that come through the

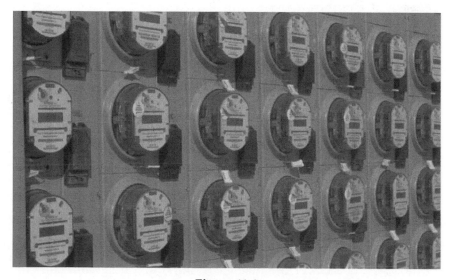

Figure 11-1.

same transformer and fed to operations that might be only a couple of feet apart.

This is how combing meters saves you money; If you were to combine your meters into fewer than you have now or even one single meter, there would be only one customer charge per month. And your usage of all the various facilities would only be run through the utility's billing formula one time.

The service level you receive from your utility will not change because of you combining meters. Many customers fear that they will get treated poorly by their utility if they try to reduce the utility's revenue, so they use that as a rationale to take no action.

Another Benefit: Reduce Demand-related Charges

When you have separate meters, each one registers a different peak demand level each month. In utility billing formulas with demand ratchets, those demand charges might stay with you for the next 11 months as a fixed charge on your bill.

Your Usage and Other Billing Factors Will Not Change Because of Combining Metering

The amount of kWhs, or energy usage, on your monthly bill will be the same regardless of how many meters you have. Unless you are coincidently reducing usage at the same time you combine the meters, do not expect any efficiencies in billing strictly because you combined the metering at a location.

Also, there are "ancillary" charges that are driven by usage and not demand, such as environmental surcharges. Those will stay the same before and after a meter combination.

How to See if Meter Combinations Will Benefit Your Company

Not all utilities allow combining meters, so ask first.

If they are okay with it, ask them to simulate what your costs would have been over the past 12 months given actual usage on all meters you are thinking about combining. Ask them also to discount the demand part of the bill by 5% and 10%.

The demand savings come from the coincident demand occurring at a different period of the month. It would be an incredible coincidence if multiple meters happened to peak every month at the same time. It is far more likely that two or more meters set their peaks at a very different time of the day and month.

With combined metering, you will only have to pay for the "coincident" peak demand or the time of the month when both loads reached their coincident maximum. This cannot be any more than the individual peaks you are getting billed for on separate meters now.

As we have talked about in other areas, getting alternate billing scenarios simulated is best initiated through an email to your rep. Such a message also gives the utility a customer communication that can be passed along to engineering.

One of the first questions your utility will likely ask you is whether the multiple meters are located on one contiguous piece of property. Does your business own the land the meters are on, plus the land connecting them? This is usually a showstopper if not.

You may discover your utility offers "multiple facility rates" that allow them to make these virtual meter combinations without all the trouble of physically or electronically combining the metering.

OPTIONS YOU MAY HAVE FOR METER COMBINATION

1. Totalizing existing meters
2. Physically combining meters
3. Metering at a higher voltage

Option 1: Physically Combing Meters

Physically combining your meters means having an electrician physically connect the two or more meters through one metering point, in concert with your utility.

Connecting the multiple electrical loads through one meter is the old-school way to combine meters, but it is most likely the cheapest. Physically combining meters gets the job done by combining meters with a one-time cost—no ongoing charges. You must have some certainty about the savings; otherwise, it is hard to take the leap and invest. Your utility can simulate those benefits for you.

You will have to get an electrician involved. Your cost will depend on the proximity of the two meters. The further apart they are, the more wiring is required.

If multiple meters are very close together, the cost to combine them could be less than $2,500. If the meters are far apart, the costs may go higher than $40,000.

You will have to pay this money out of pocket before a cent of savings is realized. But, if you are sure the operations that are metered will be there for a long time, then it may be in your best interest to bite the bullet and pay for a physical connection.

Option 2: Totalizing the Accounts

Competitive utilities often offer ways to combine the meters this way and make combining meters profitable for you from day one. This method uses electronic communications equipment to "totalize" or combine and overlap your meter data. With this approach, there will be an ongoing rental charge for the communications equipment that the utility installs.

What we are creating here is a virtual single metering point without going to the trouble and expense of physically combining them. Through totalization, all the meter data for multiple meters flows through seamlessly and creates a single picture of your usage. The utility can then use that usage profile to bill you.

Though your rep may never have heard of the concept, that doesn't mean your utility does not allow it. Also, totalizing accounts is a lot of work for the utility.

Totalizing is more the exception than the rule, and most utilities don't offer it.

The advantage is, by doing things through totalization, you will be cash-flow-positive from day one. The rental charges for the communications equipment the utility will have to install will usually show up as a separate line item on your bill.

The cost will vary based on the utility and the type of equipment they specify. Any extra charges over and above the amount allowed per facility will be rolled into systems rental charges.

Option 3: Metering at Higher Voltage

Power delivery is inherently a very inefficient process and produces a lot of waste. It is delivered to you from the generation facility over a range of voltages. But, it is often up to you to choose at which voltage you "take possession" of the power you are purchasing.

To affect the highest efficiency in the delivery of power, utilities routinely increase the voltage to transmit it over long distances. High voltage power lines are the giant ones that go down the middle of wide, clear cuts through forests and fields. High voltage is the most efficient way to move power over long distances. You cannot use the electricity at very high voltage though. The voltage must stepped-down before it is metered and finally delivered to you.

Although there will be a one-time cost involved in to relocating metering, you have a choice in where your meter is placed in that sequence of voltage reductions. You can save a lot by taking possession as far upstream as you can afford.

QUICK PRIMER ON POWER DELIVERY—
CENTRALIZED POWER GENERATION

Long ago, at the beginning of the electrification of America, the decision was made to generate the electricity in one place and distribute it to where you are and where everyone else is.

Several forces were vying for local generation very close to where people lived however. The model that won the day, however, involved generating power far out of sight of civilization and then transmitting that power over wires to where people wanted to live.

But there is a problem utilities have getting power to you. The economies of power generation are not the same as the economies of power distribution—they are the inverse. Power is generated at a relatively low voltage and then transmitted through alternating current over long distances to you at much higher voltage.

To transmit, power utilities must "step up" the voltage so it can be transmitted more efficiently.

Therefore, this is the voltage journey your electricity takes between the time it is created and the time it reaches your meter:

Generation: 1kV-30 kV (kilovolt = 1,000 volts)

Extra High Voltage Transmission: 500kV-765kV

High Voltage Transmission: 230kV-345kV

Subtransmission: 69kV-169kV

Distribution: 120V-35kV

Utilities usually bill customers at one of several optional voltage levels: transmission, primary, and secondary voltage. More than likely, your company is now billed at what the utility calls secondary voltage.

For you to save money by changing your billing to the higher voltage, the utility would need to install a different meter at a point in the electricity flow before that voltage is stepped down. The utility would remove the meters you currently have.

How Does This Strategy Save You Money

There are losses or inefficiencies associated with electricity transmission at any voltage. The electricity you use travels a long way to get to you. Many electrons that are generated never make it to you—they get dissipated or lost along the way in the form of heat and magnetic fields.

These losses come from the Joule effect. Conductors dissipate heat generated, and that primarily causes the losses.

These are the losses that are experienced at the major sections of power transmission and distribution:

1-2%   Step-up transformer from the generator to the transmission line

2-4%   Transmission line

1-2%   Step-down transformer from the transmission line to the distribution network

4-6%   Distribution network transformers and cables

The total loss of the power plant and consumers is somewhere 8 and 15%(4).

The utilities try to cut down on the transmission inefficiencies, but it is impossible to avoid them entirely. However, utilities are only going to work so hard on that problem because line losses are a pass-through item. Meaning, utilities calculate their losses and bill you for them.

In any case, by the time you pay for your energy at a secondary voltage level, you are paying a lot for that inefficiency.

By selecting a higher voltage delivery/metering point, those losses between the primary voltage and secondary voltage will be less of what you pay.

What Will It Cost?

Primary meters can be expensive (>$20,000) and your utility will make you pay for them. Also, the installation of such a meter can also be disruptive to operations, as power will need to be shut off to complete the job. Your utility will have to do the whole operation—from analysis and installation since it is their equipment. So, you pay for it, but your utility will operate and maintain it.

What Are The Risks?

The risk to you is cost. It depends on your utility. Some make it super-easy and will even rent you their upstream facilities. The ones that require you to purchase the equipment often say they will not maintain the equipment for you if it breaks.

Meters are very reliable and require little maintenance. There is not much to go wrong with transformers and related equipment, but of course, they do sometimes fail.

In the scenario where your utility is requiring you to take responsibility for the maintenance and replacement of the facilities, there are solutions that do not involve you taking on more of the actual risk. If you are required to do the ongoing maintenance yourself, there are companies you can outsource the work to. These are the very same companies often your utility outsources their work to, so they may be the ones working on your equipment anyway.

You will have to be open to creative ways of pursuing primary or higher level power purchasing to overcome some of these downsides. Your utility would prefer that you not go to any of this trouble.

By unbundling their charges and taking some responsibility for losses yourself, you are creating a much more customized pricing plan that can save you a lot.

Many power cost reduction opportunities require "unbundling." When I say unbundling, I mean looking at the real cost the utility bears bringing power to your plant. Unbundling means factoring out generalized costs. These are charges that are accrued in the service of other ratepayers that the power company is now charging you for through their rate base.

In general, it pays to research different scenarios to gain access to higher voltage power. Do not expect the power company to bring any of this up to you.

If you don't take the lead, the billing inefficiency is not going to resolve itself on its own. These costs are slow-drip eating away at your company's profitability.

CHAPTER SUMMARY

There are often opportunities to further reduce costs by renting, leasing, or buying all or part of your utility's power delivery infrastructure. There is no one-size-fits-all approach to this. Your analysis often begins with asking about the options your utility offers for taking delivery of your power at a higher voltage.

QUESTIONS TO THINK ABOUT

1. How many meters do I have on the same contiguous property?

2. What is the cost per unit of energy for each meter now?

3. What is the cost per unit of the meter with the highest volume?

4. Approximately how much could be saved if all usage could be purchased at the lowest price per unit?

5. Will the utility require a physical connect to combine the meters?

Part IV

How to Get Things Done with Your Utility

Chapter 12

Dealing with Your Utility

HOW A UTILITY WORKS
(blog.aee.net Coley Girouard 4/23/15)

Okay, now that we have a lot of avenues to reduce costs, we must realize there is a huge gap between *having* an idea and implementing it.

It is essential to understand the utility business model and how they make money before we go any further. Utilities do not think as you do as a for-profit business.

Your Utility is a Monopoly

First, your local utility is a monopoly in the truest sense of the word. They are a monopoly created by the government. The founders of this heavily-regulatory process didn't create this monopoly out of greed, but to insure reliable power and gas delivery.

Your Utility's Formula for Making Money

Total Revenue Requirement =
Rate Base × Allowed Rate of Return + Expenses

If free market competition were to create true competition between utilities, we would not be able to see the sky anymore for all the power lines. The public access areas like over-ground power poles and underground access like utility trenches would be so crowded we would not be able to dig a hole in the ground without hitting a pipe. It is well-accepted that a monopoly model is necessary to deliver electricity and natural gas and water to the public.

Utilities are therefore given the right to serve a particular market area by state public utility commissions. This means they are given the right/privilege of providing services to an area exclusively and that no competitor can come in and take away that business from them. States control incumbent utilities tightly. This situation also creates an incumbent or monopoly mindset. Including attracting very conservative risk-averse employees. Unfortunately, these are people who often only marginally understand your need to make a profit in a competitive business environment.

How Your Utility is Carefully Regulated

State public service commissions provide constant oversight of public utilities. The level of oversight all depends on the state in which the utility operates. The state public service commission must approve any changes requested in the way your utility manages or bills you.

Allowed Rate of Return

Because of the nature of a monopoly, the state public service commissions all regulate the way utilities can charge their customers. The consumer of the monopoly service has no option but to buy from the monopoly supplier.

Therefore, utilities are granted an allowed rate of return on their investment or, assets under management. They spend whatever money is necessary to create and deliver power, and their expenses are simply passed on to consumers.

This is a system that would be abused without statutory oversight. State public service commissions carefully watch to make sure utilities do not spend too much money.

The allowed rates of return vary greatly depending on the state. Alabama Power Company has traditionally had the highest return at over 13%, but the average is closer to 10%.

Your Utility's Expenses Are Passed on to You

A utility's expenses are 100% passed along to its ratepayers.

So, if left unchecked, utilities are incentivized to spend a lot of money and send a bill for those "costs of business" to consumers in the form of increased rates.

Fuel costs to run generators are a particular area where sloth can drive up costs. There is no incentive for your utility to do a great job at managing those costs. Utilities are not incentivized to keep fuel costs down because fuel is simply seen as another cost of doing business that can be billed to someone else.

I am not saying these things to accuse or even complain, but it is important for you to know how the utility mindset and system is different than those of a for-profit business.

Rate Cases

Rate cases seem always to make a case for more money not less. They allow utilities to petition their state public utilities commission to increase their allowable asset base and rate of return. Everyone with a voice in the matter is encouraged to weigh in.

Utility Mindset

It is important to note that very conservative people staff utilities. Working at a utility is a very stable platform to work within. Layoffs are seldom because as we've talked about, they rebill their expenses to the public in the form of increased rates so there is never a day when they run out of money as a traditional for-profit business can.

So, for example, if you bring up that you would like to consider interruptible serviced, do not be surprised if your utility rep suggests that you not do it. They very often only have an on-paper understanding of their own rates and not a practical understanding.

Rates are legal documents, and they read like legal documents. Especially the interruptible rates. They spell out penalties and worst-case scenarios. These documents are written to protect the utility from liability. Rates are a nightmare of jargon and confusing terms.

We must realize utilities have a higher responsibility to the other utilities they interconnect with than the customers they serve. That means they must do whatever it takes correct any imbalances that might affect neighboring utilities. This usually means, shut off some load on their system if they must keep the grid intact.

If they don't shut down some load in such a circumstance, they run the risk of that imbalance spreading across multiple utilities and causing involuntary brown/blackouts. So, that means, if a utility feels your load needs to go offline for any reason, they will take you offline.

This is an important perspective to have when considering non-firm rate programs. At least if you are on an interruptible program, you will get paid for the inconvenience and hassle. The companies that haven't signed up for an interruptible program or taken on demand response programs may be taken offline without compensation.

As we've discussed, the key is to know how close you are to the precipice of an involuntary interruption anyway, without the interruptible power program in place.

Having tenacity is something that will help in dealing with utilities. Many representatives aren't used to answering detailed questions, and they usually don't readily have the answers. Since such questions aren't part of their normal workflow, they may have to re-pose them to others who won't get back fast or at all. They may also get frustrated that you are asking these questions in the first place.

There is gold at the end of this rainbow for you, but you will have to sort things out for yourself. This part can't be subbed-out to the utility. Remember, they have little incentive to help you reduce the revenue you pay them.

How Your Utility Bills You

If you are like most people, your understanding of how your utility bills you are simple—they bill monthly and you pay it. There is more to it than that. Having a working knowledge of which factors on the bill mean something will help a lot in understanding what you can do to your lower costs.

You must get on your utility's side of the table first and see things from their perspective. Their billing models reflect their economies of scale in producing power and their desire to market to new industries.

The first thing the utility must do is cover their operating costs. The economies of running a power generation facility require that the facility is run 24/7/365. Since industrial companies often run with that same production schedule much of your utility's rates for business will usually incentivize for businesses with the same heavily off-peak run schedule.

As far as rates are concerned, utilities care about four major components of your consumption when developing your invoice:

1. Immediate demand
2. Average demand
3. Usage over time
4. Power factor

Your instantaneous and average demands are the main things we are concerned about, since it equals about half your bill. When we are talking about "demand," we mean the amount of energy you are using at any one moment in time. Over time, the collection of immediate demands are compiled to form your usage or kWh.

The reason your demand readings are important is they are something you can control, and the utility also has a real interest in your controlling them. Most utility programs incentivize you to back off on your demand level at different times of the day or different seasons of the year.

This reflects the strains on their system as well as their interchange agreements with adjoining utilities. A thorough understanding of the underlying drivers moving your utility to price demand the way they do will illuminate a wealth of opportunities to save money.

Now, most utilities offer programs to incentivize you to reduce demand load during the utility's peak times of output. For most, that means during the summer months. Since air conditioning and the need for climate control occurs at the same time as when large industrials, commercial buildings, and government buildings are operating at near peak demand, you can see the need for greater control over demand during the summer period.

So, if you dig a bit, you will find that your utility will offer programs that can significantly cut your costs by mitigating summer demand. This can require adjusting operations to take advantage of incentives. (It may even be lucrative enough to consider passing on some of the savings you receive to the employees who shift their personal schedules to accommodate the better utility rate.)

Another class of utility rates that rewards demand control is time of use. These billing algorithms stratify incentives and encourage your business to back off spiking demand for two time periods: summer/non-summer and during the day, stratified between peak and non-peak.

The bottom line is that your utility cares a lot about your demand level and they can incentivize you to modify your demand at different times of the day and different times of the year.

You may have to go upstream of the utility to get the answers you need. The solution to cost reduction may have to be found upstream of your utility with *their* supplier. Their supplier will often be a much larger group of generation facilities. They will have a contract agreement with that entity and inside that agreement will be the solution to your cost reduction problem.

In other words, the arrangement they have negotiated with their supplier will likely have incentives that can be passed down to you. You won't know unless you ask. You may have to start the dialogue with them through your local utility though.

Your utility's supplier may not talk with you directly unless your utility is at the table and authorizing them to do so.

Often you will find that your utility's supplier will have worked out rates and additional ways to save with other utilities they supply energy to as well. These programs with other utilities can often be activated by your utility for your benefit. You must know the right questions to ask to access this information.

Why Would My Utility Do Something to Help Me Reduce The Amount of Money I Pay Them?

I get asked this all the time. That is because, from a business person's perspective, it doesn't make sense for them to help you very much. The utility gets nothing out of it. Since they merely rebill their expenses to you as the consumer of their services, they are just as incentivized to do nothing. Helping you reduce the money you pay them makes no sense outside of this government mandated system.

What Motivates a Utility?

The good news is the same things motivate all utilities so once you understand this, you can apply it anywhere.

What leverage could you possibly have? As it turns out, you have a lot of leverage.

First, there is a big difference between the goals that your utility has and the goals your local rep who has been assigned to you has. While your utility's management is very revenue driven, your local rep wants to make you a happy customer. Your local rep isn't motivated by revenue so much but rather by achieving a high score by you on customer service every year. You will find that most reps will bend over backward to help you for that reason.

Secondly, if you are an expanding operation, you increase the money you pay them. While you don't have a choice from which you buy power from where you are now, you do have a choice where you build additional capacity and who you buy that capacity from. You can locate new equipment in operations and equipment in other utilities' service territories. This is a very important point of leverage you have as a manufacturer.

You will receive a much higher level of customer service by the "new account acquisition team" than you will by your regular customer service people. When you activate this team, you will have the opportunity not only to reduce any future costs for the new load you are considering, but also, you will likely have the chance to reduce the cost of the power you currently purchase.

While you do not want to mislead anyone, of course, you need to speak up about potentially expanding load. If you are considering new load or a new production line, you need to let your utility know about it in detail as well as the options you are considering. This will put you in a much stronger position. The utility's most creatively entrepreneurial people will be brought to bear on the "problem" of you having additional electric or gas load that needs an incentive to locate inside their territory and not elsewhere.

I have seen utilities get into a pricing war with each other over a potential new customer. Utilities will become very competitive and quickly drop their prices if they feel the new customer is worth acquiring. In such a competitive context, they will have way more power to get creative and reduce your costs so it pays to get speculative with them about any new operations you may add in the future.

HOW TO WORK WITH A UTILITY MOST EFFECTIVELY

How, then, should we approach them about cost savings? The best way is to be specific in your requests. Specific simply means asking questions that cannot be answered with yes or no. It means asking for exactly what you want to know.

In rate analysis, that means asking your rep to simulate the impact of any possible rate changes and work with you to get answers. Once you get any analytical reports back, do not be intimidated about asking more questions and asking your rep to do some more work. If you would like different scenarios simulated, ask them to do that. And, challenge any variables they have pre-loaded into your scenarios that you don't think apply.

An engineer at the utility will do most of the analysis you ask for. Engineers at utilities are overworked. Your utility rep may be protective of their engineer's time. We can feel sorry for them but here is where you need to be a little selfish. You are entitled to this information, and you need it to lower your costs.

I have had many a rep tell me a billing scenario is hopelessly flawed and then find out it saves the customer 10%. Be persistent.

If You Are Working with a Consultant...

It is best to keep the communication between only you and your utility. Fight the urge to pass communications off to the consultant. Your consultant talking with them will usually only slow things down. The utility and their rep may see the consultant's participation as a threat and that "you don't need them—I can do anything they can do for you."

That is not at all true. There wouldn't be consultants out there if utility reps were empowered to help their customers save money.

How to Get Things Done

You do not need to nail down all your questions before opening up a dialogue on savings with your utility. You need to know what you want to achieve and that is cost reduction. Although it may seem obvious to you, this goal needs to be articulated to your utility because it will not be obvious to them.

Your attitude should be that this is not a casual exercise and you will be doing whatever it takes to ferret out options.

Many times, you will have had to politely go to reps and managers of reps repeatedly to get answers. Do not be afraid to do this and do not be afraid that your utility will somehow penalize you for speaking up. While this kind of thinking is what stops many ratepayers, nothing could be further from the truth. I have never found anyone in that world who would penalize you for asking questions.

If your utility feels your heart isn't in it, however, they will treat you accordingly. They will give you ultra-high-level, back of the envelope answers and hope you just forget about it. Utilities are cutting support staffing to the bone, and reps are very busy. Your utility rep is not seeking out new projects from you to entertain himself.

I worked with a large electric utility a while ago that had an entire team of people, including some PhDs, dedicated to discouraging customers from building their own power generation. While it makes sense that an electric utility would not try to convince you to make your own power, any advice coming from your utility of a savings nature must be taken with a healthy helping of skepticism.

Utilities can be a good source of the information you seek to reduce costs. But you must craft your questions well and aim them at the right person on the organizational chart. Utilities can assist you in lowering your costs if you know beforehand what you are trying to achieve and you are persistent in making sure you get the information you are looking for.

So, craft your questions well. Instead of, "Am I on the best rate?" the question needs to be, "Can you run a comparison of all the rate options available to me, based on my last 12 months of usage?" That is a question your rep can work with.

Always ask for an expected time for them to get answers back to you. I have seen the simplest of questions enter a black hole and take months to get turned around.

Nevertheless, it is important to treat your utility and their representative as partners. The culture is oriented to help you more if you act as if you are in partnership with them. Look at them like they are resources in your quest to save money. They can help you, but they are not going to be as relentless as you in pursuit of savings.

CHAPTER SUMMARY

Your utility has no problem with reducing your costs if you can push to find the savings. Do not expect them to be proactive, though. The best way to approach your utility with any request is to be very specific. Not, "Are we paying the lowest price we can for power?"

QUESTIONS TO THINK ABOUT

1. Has the utility assigned a representative to work with you on our accounts?

2. Have you let the utility representatives know you want to work with them to reduce your costs?

3. Have you adjusted your expectations and internal metronome to "utility time?"

4. Have you identified others at the utility I could appeal to if things stall working through the "normal channels?"

5. Do you have the option to purchase service from another provider?

Chapter 13

Appoint a Champion and Get Things Done

"The credit belongs to the man who is in the arena."
—Teddy Roosevelt

Figure 13-1.

WHY PEOPLE HAVE TROUBLE IMPLEMENTING THINGS

We are in a state of Nirvana when we are analyzing and before we start the hard job of implementing. Everything is in the future, and everything seems promising. It feels like we are doing something significant but without any risk to our careers. It is easy to stay in this analytical state and never leave—many never do. There is always a little more that can be learned, a little more to be studied. Projects can drag on for years moving from 95% to 98% certainty.

Even if our organizations give us license to move forward, it is hard to muster the will to take final actions and make things happen. Without a mandate, we are lulled into inaction.

The ideas we have presented are new. There will be real pressure from people inside your organization to stick with the status quo and take no action. "Great work. This is not for us though. Let's get back to what we do best—making widgets."

There are high-minded sounding phrases too that will undermine your best efforts to change things. You have heard them all. Our projects are often at the mercy of smart people who want to kill projects so they can get things around here back to normal.

- "Stick with your core competencies."
- "We are not in the business of saving energy. We are in the business of making widgets."

Commonsensical but spurious, these phrases are enemies of progress.

From the outside looking in, plants often seem to be the picture of unity. But, from the inside looking in, they are often acrimonious. The reasons are many, and this type of culture is widespread in manufacturing. Almost all plants display a quasi-hostility towards real change.

People in the maintenance function of a manufacturing facility are charged with keeping things up and running. Anything from the outside that challenges their ability to control uptime will meet with resistance. Not only that, but a lot people who work at a facility all day become very territorial.

Have you ever presented a new idea for improving your company? Chances are you got turned down or maybe even laughed at. That is manufacturing company culture so try not to take it personally. There is a deep risk aversion.

We humans seek to limit risk. We all do it. The problem we must recognize is that company cultures like this tend to intensify. They get stronger and stronger and more territorial over time. This is not something you are likely to change, so plan for it within the context of your project.

That means mentally charting out the likely attitudes and responses of the people you are likely going to need buy-in from within your organization before you begin. Some people want to appear to go along with new ideas because they think that is what upper-management expects from them. They pay lip-service to the idea and never intend on supporting you, or even worse, they are hoping the natural entropy not to change things will kick in and your project will collapse from its own weight.

This is not being negative, this is the manufacturing landscape, and it must be dealt with. I have been in at least 1000 different manufacturing operations in my lifetime, and the culture of risk aversion is alive and well in almost all of them. We can be polite about what we call it, but the net result is an aversion to any change. The founders of your company were forced to be entrepreneurial and look seriously at all new ideas to find hidden savings. That is how they survived. They could not afford such an attitude of complacency when it came to new ideas for cost reduction.

But the culture that created your company is not the same culture that maintains it now. We must adapt to the way things are though and forget about trying to bend our company culture to fit us or our project. There is too much inertia.

You may be thinking; "Why are you making such a big deal out of this? My plant maintenance people grumble a little, but they do a great job for us. They keep our plant running through thick and thin and without them, we would miss product orders all the time."

This is very true but to survive and thrive as a business, you will have to take limited risks on or you will not keep up with your competition.

In this sense, the project champion you select will be providing this oversight and channeling the company founder's attitude of risk assessment and actuation.

This is the most challenging subject we will discuss because it involves getting things done through other people. It doesn't matter how brilliant your idea is or how much money it will save your company; it will not fly if someone does not drive it and make it happen.

Always appoint a project champion. Without a person chosen who is responsible for success, projects are too prone to fail under the weight of the status quo.

Your project champion will need to be a bit of a salesman. They need to be optimistic. Regardless of the strength of the project, complicated dynamics inside most plants are lurking to kill efforts to change things and improve.

The recommendations for cost reduction we are talking about in this book will be perceived differently by different people in the manufacturing process. For example, since people who work in maintenance are the folks charged with keeping the plant up and running, they will have no interest in considering interruptible power or anything that could make their jobs more difficult. A potential interruption is seen as a hassle.

We all protect our own turf and try to minimize our own perceived hassles. That is why having a champion on watch is so important. A project champion can take the wishes of management to reduce costs and translate those into terms each faction within the organization can understand and live with.

Maintenance must be shown the perspective of management by the champion and realize that the service that is being pursued to cut costs will not negatively impact them. Maintenance staff is very much needed to implement many of the cost strategies I am recommending. If there were to be a power or gas curtailment, for example, Maintenance would be integrally involved both in shutting the operation down, maintaining it during the shutdown, and in eventually restoring the plant to full production capacity.

WHO SHOULD YOU CHOOSE TO BE YOUR CHAMPION?

Figure 13-2.

The person you recruit to be your champion should be someone who gets things done and who has a propensity to do things for their own reasons, not relying on the opinions of others to tell right from wrong. This person cannot be easily swayed or someone who goes with the flow. Remember, the "flow" has gotten you where you are now.

The primary directive of the project champion is to run interference. To work through the details of the program with others in the organization, handle the speed bumps and get things done; they will have to fend off the naysayers which abound.

While they come by their fearfulness honestly, well-meaning bureaucrats can make an aggressive attack against the most obvious and sensible of cost reduction efforts. If it wasn't created by them, it has no value. It is new. It hasn't been done before. "That's not the way we do things around here." You've heard them all before.

As we've said, the people whose job it is to keep the equipment running in the plant like to eliminate as many variables that may stand in the way of continuous uptime as possible. Of course, 99% of the time, you want them to be wired this way. Their protective nature is what gets them out of bed in the middle of the night to fix machines and keep product moving out the door. While this personality type is essential for steady-state operations, it is the kiss of death for cost-reductions efforts.

Your champion will need to be perceived as having authority by all involved. Therefore, the best person to choose for the role will be one both sides get along with and respect.

Before the advent of consensus-style management, plant managers usually made these decisions without much input from the floor. There was no discussion, no debate, and no one cared whether plant personnel liked a decision or not. Now, plants subscribe to a leadership style where everyone is encouraged to participate in decisions. While this can often lead to better decisions, it also just as often stifles progress.

This style often leads to the strongest personality or the strongest department prevailing, as opposed to the strongest idea.

The cost to the company of honoring personalities over results is enormous. The roadblocks are not always broken by this liaison style of management. It is important to consider the choice of champion carefully.

PROJECTS MANAGED THE NAVY SEAL WAY

One of the U.S. Navy SEAL's favorite phrases is, "you win more when you act with boldness." Once we have analyzed things and we find opportunity, we owe it to ourselves and to our companies to act swiftly.

Unchecked, our minds can be fearful and lazy. They want us to do nothing and delay things. "We need a bit more analysis" gives our minds an excuse to sit tight and take no action.

This is very easy to let happen. There is always more data to collect and always more thinking to be done.

So, we need a single project champion to keep us on course.

THE TRUE VALUE OF THIRD-PARTY OVERSIGHT

Left to their own devices, people often put decisions off. People are stretched to the limit and companies do not have the internal resources to dedicate one person to simply oversee another.

Figure 13-3.

Without someone overlooking us most of us revert to not taking action. Why? Because there is risk associated with taking action. It may be a very small risk, but it falls under the rubrik of "risk." You are usually moving from a position of certainty to uncertainty. Even if the position we find ourselves in now is not such a good one, at least it is one we are familiar with.

The only way to effectively combat this dilemma is through the power of a third-party observer or champion. Without the pressure of compliance and time constraints that champion brings to the table, things do not happen. Champions can push and will be judged by deadlines met.

There will be some who challenge the authority of your chosen project champion. For those who insist on bucking the charge of the champion to manage the details of the project, there are remedies.

Having a Single Individual Be Responsible Has Power

Having a single person that management can look to for project status and project makes things much easier to control.

The proviso is the project objective must be made crystal clear to all involved. Without this kind of clarity, there is no way you can hold the individual you pick for the champion's role to any standard or objective measure of success.

The Best Leaders Do Not Always Have a Title

Where you pick your champions from can have a large impact on the outcome of the project. Obviously, your top candidate must have earned the respect of the people they will be leading. You cannot pick the new engineer right off the cabbage truck to be your champion. That is not fair to your project, or to the new engineer.

Tools the Champion Must Use to Move Things Along

The champion of your project must first commit to a timeline for completion. Many people bristle at the thought of committing to things, especially to time constraints. They often feel like it is too difficult to say when things will happen because there are too many variables, most of which are outside of their control.

A timeline is crucial for success, though, and here is why that is the case: Without a guide for the time dimension, things get lost or sidetracked in favor of other goals and other objectives that are more concrete and definitive.

One reason we default to doing our C-items over our A-items in life is those items are more concrete. We can envision them easier, and the C's give us a warm feeling because they are clear and we know we can do them.

All large tasks must be broken down into C-items for them

to get accomplished. The problem here and the reason so many of our goals stay in the planning stages and eventually fade away altogether is we never take the time to pull the C-items out of the A-list or the big project. Why do we do that when we agree upfront that the A-list items must be done?

The reason is the detailed breakdown of a giant project into small steps is laborious and involves a lot of estimating. Unless you can envision and gain a concrete picture of the eventual outcome through this breaking down process into C-items, you or your champion will never make progress. Because management of the details and pushing even when it feels like we are making no progress is the special sauce we need to get things done.

It is usually best to let people come up with their own timeline and their own list of tasks to complete. Even if is not up to your standards, you must insist that your champion plan in detail how and when things will get done themselves. With that personal planning, people develop buy-in. Buy-in will generate the commitment, and the commitment is necessary for the success of any project.

If your chosen champion has worked on a lot of completed projects, he or she probably already understands that. But understanding the concept is often very different than doing it.

But regardless of the level of experience of the person you pick to lead the charge and be your champion, we are all wired the same when it comes to getting things done. You will serve the champion and your facility the best by forcing them to get clear on the micro-tasks needed to accomplish what they have been called on to accomplish.

Going into a ridiculous amount of detail in planning should be encouraged. Counsel your champion to embrace this process for the mental cobwebs that the exercise clears away. Make sure they realize that you realize detail and action steps will change and it is highly unlikely that the C-items and timelines they create will happen as planned. Give them confidence in the strength of the goal-setting process through putting down a detailed timeline. Make sure they realize that you aren't trying to create a stone tablet for Moses to carry down from the mountain. The timeline they create for you will be updated weekly or even daily if there is a lot to do with your project.

I once worked with a very successful consultant who encouraged me to write down a massive list of everything that could derail a project or task. Then, he said to go about eliminating those hurdles systematically, and the accomplishment of the objective would happen on its own.

While that could be construed as a negative way to look at things, it worked. By process of elimination, things got done. Michelangelo chipping away at the stone and eliminating what was *not* the David created the statue that *was* David.

It is not only ok but probably also desirable that projects start out chaotic. If the steps were clear, it would not require high-level intervention to solve. If it were easy and obvious, it would have already been accomplished. You and your champion are the ones who will shine a light along the path, bringing clarity where there is confusion, and accomplish whatever it is you set out to accomplish by your commitment. Eventually, turning A-level thought into C-level reality.

If you and your champion can become comfortable in this land of uncertainty, you can accomplish anything. The results you can achieve are not limited to the energy genre. These project execution skills will transfer to any project you choose to tackle in the facility or your life. Being ok with not feeling under control is a necessary component of success no matter what the area of endeavor. Striving for something that may seem way too big at first is the grist of champions.

How to Know if it is Time to Shift Horses Midstream

It is sometimes the case that you will have to pull the person you initially picked to lead the charge and replace him or her with someone else.

The only way to find out if someone is great at pushing through and getting things done, though, is by letting them try. That often means you find you have picked the wrong person.

Either their personality does not match up with the people who must be led, or the details overwhelm them. Changing your champion and moving to a new project leader sometimes must be done to keep things moving. If things are dragging, changing to another leader is the right thing to do.

In this way, picking a project leader is not unlike picking a college football player who a scout hopes will do well in the pros. Stories abound of superstar college athletes who everyone thought would translate their magic into the pros and they flopped. We all know of the Super Bowl MVPs who the scouts never thought would amount to anything. There is no way to know how well they will do until you put them on stage. Some people rise to the occasion, and some do not. The only way to find out is through trial and error.

In any case, react decisively and remove the weak project leader fast before their lack of effectiveness can derail your project.

This, of course, is all within the context of an agreed upon timeline and action steps. There is no good way other than by this objective standard to judge someone's work.

All relationships in life and business boil down to setting expectations and frequently communicating about whether those expectations are being met.

CHAPTER SUMMARY

Many forces resist internal attempts to change the status quo in manufacturing plants. Therefore, it is essential to appoint an internal champion to lead the charge on an energy project if you expect to get anything done.

QUESTIONS TO THINK ABOUT

1. Have clear objectives been set? Describe in writing the end state you want to achieve, for example, "Implementation of

demand response in the Houston plant before 12/31."

2.   Have you identified at least 3 individuals as candidates who have the ability to work with the key players at your company?

3.   Once the project champion is chosen, have you explained his/her authority to the people the champion will be working with?

4.   How will you measure progress?

5.   How often will status updates be required and in what format?

Chapter 14

Long-term Monitoring and Acting on Your Data

"If there is no struggle, there is no progress."
—Frederick Douglas

You must be objective about what you learn from energy data. What sources of waste can be mined here?" Have a bias for taking action and investigating further.

Once your energy management system is in place, your task is to set up a regular reporting system to keep the data front and center. This will create a virtual perpetual motion machine of continuous improvement.

All of us want to improve, but we need feedback to know whether our efforts are bearing any fruit. By setting up a simple system of reporting energy data to those who can use it, you are giving them the tools to solve their own problems. By giving multiple people in the organization the raw data, you are creating social pressure to use that data to reduce energy costs.

What information and conclusions can be made from raw energy data? Waste for one. Any energy used that is on your utility invoice, but is not necessary for getting your product out the door is wasteful.

Wasted energy has been estimated by the IEA to constitute at least 25% of world energy production. That means that 25% of the world's energy generation is running merely to supply power that is unnecessary. This may be a conservative estimate. Next time you are driving around town in the middle of the day, notice how many parking lot lights you see on. Some burn round the clock. This kind of waste that is glaringly obvious is a sure indicator that there is far more waste lurking beneath the surface.

Most is esoteric and is a case of out of sight, out of mind. We only tend to only give attention to the alligators nipping at our heels. There are studies you've probably seen where feedback is given to drivers on an electronic sign regarding their current speed. Seeing their speed, drivers slow down. It doesn't matter what the speed is; they hit the brakes. Awareness has an enormous dampening effect. The more visible and real-time the notification, the more powerful the effect.

The situation is no different in a manufacturing plant. When we give energy data to those using the electricity, we need to make sure that it is in a public setting to be the most effective. We can thus praise those who work to reduce energy waste and introduce some social pressure to motivate people and encourage the actions we want.

Most of us want to look like a responsible member of the company to the fellow members of our group. Self-actuation makes us want to show ourselves that we can take actions to improve. By letting the broader group know what individuals are doing, we are creating an environment that can lead to improvement. It will become automatic where the players involved are actively working to continually reduce usage without a manager having to do anything other than making sure the data is being delivered and visible.

Successful energy management is about continual improvement and seeking out the waste hiding in the cracks. Once you have initially let the EMS system point out the obvious big-ticket items to fix, the battle is only half won though. Ongoing improvements can often dwarf the initial set of improvements.

HOW DO YOU ACCOMPLISH BIG THINGS?

Stop "Death by Analysis"
Being an engineer, I understand over analyzing things. When you get an assignment to study something, it is very easy

for the analytical side to get out of hand. Why is that? The main reason is it avoids judgment day. If we continue to analyze, judgment by others is postponed. Analyzing offers the nirvana of promise, of working on something with potential. It gives a feeling of self-importance because we are doing great work.

Are Our Conclusions Correct?

These all are questions that bounce around in our subconscious while we are working on projects. The way out of this analytical valley of death is to set time limits and force yourself to stick to them. Set time limits early on. Make your deadlines very specific as to date and time. Discipline yourself to follow through on what you first set out to do.

If you don't set hard deadlines for yourself, the mind will always rationalize a way to keep the analysis going.

The human mind is incredible to behold as it wriggles out of work and squirms to get out of making changes. It never ceases to amaze me the rationalizations my mind can come up with to keep me from doing the simplest things. These are even things that clearly benefit me.

At the time of my self-imposed deadline, this is especially so. My brain will tell me things like; You can do it later when the timing is better and the best of all—I don't "feel like" doing it right now so let's do it later.

So, part of our problem with wrapping things up is the feeling that "this is it, there is no turning back." We cannot afford to look at things that way. No matter what the naysayers would have you believe, there is always a way to revive a project regardless of how bad the initial reviews.

We cannot afford to hold onto the attitude that with another 1% effort and we can win. Usually, 90% is good enough. Continuing to analyze at that point does not typically produce noticeably better results.

This book is about energy, but it is also about getting things done which I consider the unexposed part of the iceberg. People tend to quit in the middle of these types of projects.

The reason people quit in the very middle of energy projects is the same reason people quit anything worthwhile right in the middle. The middle of any project is the point of no return—it is midway between starting and throwing a party. The party is too far away yet to motivate anybody. So, all you see is how far you have come and how little you have to show for it. It looks like it will never work. Rewards, however, tend not to be linear. They come only at the end.

It is very easy to give up. "Why was I thinking I could do something this big?" Or, "What was I thinking I could get something done at this plant and with these people?" It is easy to be down on yourself and abandon your efforts.

There is a solution. It is not an easy one. You've probably heard it before, but you are most likely not doing it.

The solution is to celebrate with tiny victories along the way. Look at your next step, not at the miles of steps you still have ahead of you. This makes achieving a goal a much more manageable effort. Daily victories are always available.

You must start with an overview and then break down each major step into four or five tinier steps. Your mind has an easier time grappling with smaller steps

This planning phase is crucial because it is going to give you the confidence to work on tiny daily actions. Putting together a long string of actions aligned with a goal is the secret of all successes.

It is far better to spend time on the front end designing your task list than to figure things out on the fly. The pressures of your busy day will keep you from moving forward if you have not made the path before clear. There are too many random requests that drop in from the sky. It is unlikely that you will accomplish anything if you wait until today to decide what you are going to do today.

ACTION PLANS

- Get the overview right first—"Finish before you start." Plan thoroughly

- Focus on micro-tasks (less than 2 minutes)

- Accept that you will have to make changes along the way

- Commit to never quit

But far more important than any of these steps is creating the impulse to act on your intention. We are all filled with great ideas. We know the right things to do. We know how to do them even. However, we take no action because we lack the impulse to act.

But, how can we make ourselves do what we know needs to be done?

5-Second Rule

My favorite technique for getting off home plate is Mel Robbins's 5-Second Rule. Our minds talk us out of most things if we let them. Mel says we have 5 seconds before the rationalizing brain takes over.

Imagine a 5-second countdown timer that you start every time you want to take action. Instead of worrying about doing the task, concentrate on starting the imaginary 5-second timer. This shuts down your impulse to rationalize. That is because your mind is busy counting down.

Commit to making energy data available daily. Use these and other simple tools to overcome inaction on what the data are telling you.

CHAPTER SUMMARY

Self-doubt and procrastination are inevitable with new ventures. It is easy to abandon these efforts for more pressing issues. By making the reporting of energy numbers central, you can drive people directly involved in the day-to-day process to continue to cut costs.

QUESTIONS TO THINK ABOUT

1. What energy data are available in my facility now?

2. Which of those available data points are controllable?

3. Who in the organization can most easily control and improve those data points?

4. How can that data be made available to them close to real-time?

5. How often should we review progress?

Chapter 15

Summary and
Plan Going Forward

Now it is time to make things happen.

We have come a long way since thinking there's nothing that can be done to lower utility costs. Much of what we have discussed is nothing more than aggressive problem solving applied to the utilities. But, to have the confidence we need a roadmap and the assurance that comes from others saying what we would like to attempt is possible.

THE CHALLENGES AHEAD

The path ahead will not be an easy one, or you would have already gone down it. But the rewards for tenacity are great. There are many within manufacturing companies who believe that leaving well enough alone is the secret to their success. Those people seem to be a part of all projects.

KEY POINTS OF THE NEW APPROACH

There's a lot we can do to cut utility costs in a manufacturing plant. But it takes a different approach than going to lunch with your utility rep hoping *he or she* will figure it out for you.

• You must do your own deep research

- Collect options

- Assess your *current* risk exposure

- Assess the incremental risk exposure inherent in the alternative programs

- Plan how to address the naysayers

Installing an EMS Keeps the Savings Ball Rolling

Monitor your usage. Having a window into your energy spend daily forces you and your team to think about it, which will naturally result in waste being cut out.

The energy management system can and should be your cost reduction metronome. You can use the data it produces to generate discussion around areas of waste and potential improvement.

The Power of Doing Things Now

Napoleon said about time, "once lost, it is lost forever." In energy cost reduction, we cannot escape the importance of getting things done as soon as possible. Energy is not visible, so "leaks" are not obvious.

What's at Stake Here?

Thousands, if not millions, of dollars are in the balance for your company. And every day that you put off presenting your findings and launching the venture, the more money is lost, and that money is unrecoverable and gone forever.

Can-do Attitude

It may sound trite but if you have a positive, solution-oriented attitude when evaluating data, you are evaluating. You can always justify a "no" if you have that frame of mind. You cannot even look at these issues dispassionately and expect them to carry their own weight without your motivated voice. You must be enthusiastic about the possibilities.

ANCILLARY BENEFITS OF ENERGY SAVING

As a side benefit of energy cost savings projects, change begets change. Focusing on this one area can trigger a chain of improvements throughout your organization as people are motivated by your success. The momentum that a successful cost reduction effort can have on an organization can be tremendous if success stories are publicized internally.

It All Begins with Belief

It is often not possible to see or even correlate the actions we take today with future realities.

It is very challenging to correlate the actions we take today with faith in future outcomes. That is the big reason why short-term change is so hard for everyone. Confidence is hard to come by when the results of our actions are likely far out in the future.

That is why seeing the results others have achieved is so vital. You must believe project results are achievable before you will persevere and fight for them.

Risk Analysis

There is no magic to risk analysis. There are six simple steps:

Step 1: Identify the hazards. To identify hazards, you need to understand the difference between a "hazard" and "risk." This means understanding your "baseline" risk. The words we use to describe what could happen in the future are critical. The word risk has a negative connotation but, in fact, only has relative meaning.

Step 2: What is the worst-case scenario and how could the business react if the worst happened. If the worst-case scenario does happen, is it recoverable?

Step 3: Decide on measures to control or limit any damage from the worst-case scenario occurring. This means understanding the difference between what might theoretically happen and what more than likely will happen. You cannot afford to insure against everything.

Step 4: Record what you find out through your research. It helps to clarify your thinking and expose holes in your logic to write everything out.

Step 5: Review your assessment and update it often.

Step 6. Be bold. There is great power in taking steps forward.

Risk is a Four-letter Word

When an action or project is described as "risky," it sends people shivering and huddling together under the nearest table. No one wants to be near a project or action that has been labeled "risky." Once this bell is rung, it is very difficult to un-ring it.

Let's face it, no one wants to be wrong. The way we often think we avoid being wrong is not to act. But protecting ourselves in this way is no victory. Manufacturers need to have a clear sense of what is possible and struggle hard to learn how to achieve it.

Every action we take affects the future and involves a degree of risk. Every decision can be perceived as risky if we allow the wrong people to frame things. I have seen many simple projects that would save a company millions slide off the rails because of a single naysayer calling a project "too risky."

Doing nothing carries risk, and this risk needs to be documented. It may seem like the safest action to do nothing, but it seldom works out that way.

Think about the clarion call of the securities industry—"past performance does not indicate future results." Security analysts always add that disclaimer because the status quo is usually just not a secure haven in a dynamically changing environment that you have no control over.

FINAL THOUGHTS

The lesson of this book is that substantial energy savings are possible for manufacturers. It is important for you to dig for details and do your own research. And dig for them far deeper than you are used to in order to get a firm grip on the relative risk they present. This level of due diligence and risk analysis is where the energy gold is buried.

The second and equally valuable lesson is to set time limits on your work and hold yourself strictly to them otherwise analysis can go on forever. The work will always expand the time available.

There will be people from within and from outside of your company who will throw rocks at your efforts.

Lack of clarity kills energy projects. When people do not have a clear yes, they always say no. We all strive to mitigate perceived risk in all we do. Gut reactions to data and circumstances often suffice as good enough.

But, the status quo is no refuge. People in manufacturing companies often feel this is a safer position to be in, but it is not. Energy prices always to up.

The actions I am recommending are not just a nice thing to do; they are necessary for the survival of a manufacturer. The incoming tide of energy price increases is relentless. It is important to aggressively fight them with countermeasures such as those presented in this book.

Quick-start Checklist for Taking Action that Saves Big Money!

1. Call your electric utility. Ask about incentives your utility has for reducing your demand.

2. Thoroughly understand the risk and reward of other options and your current situation.

3. Be the project champion or appoint one.

4. Ask your utility for the next steps.

5. Invest in an inexpensive and simple energy management system.

6. Report usage information to those who can use it to make changes.

You will be successful at lowering your costs substantially if you follow these procedures. It will not be easy. People will work overtime to shoot down projects like these, both inside and outside your company. Do not give up. You are not the first to have run the gauntlet. This race goes to the persistent.

Addendum

HOW HAVING "DSIRE" CAN SAVE YOU MONEY

Our energy prices are going up, not down. You must look out for your own plant to aggressively fight against the rising tide. No one is going to make you do this. If you don't aggressively advocate for lowering your own costs, your prices will slowly go up, and profit margins will slowly deteriorate.

The landscape changes constantly regarding rebates and incentives provided by the federal government, state, and local officials. Every municipal entity has their own spin on how to incentivize you to save.

In some places, you can step across the street and get a different program for reducing costs provided by the utility next door. It is very confusing without help to figure it out.

How Can DSIRE Provide Help?

This is the closest thing to a clearinghouse for energy programs that we have. The DSIRE database contains all the information related to energy rebates and programs in the United States. Regardless of where you are located, this website can provide you with the most up-to-date information into what is available to help fund any capital outlays that may be required. And it's free!

It is far better to use this database than it is to do individual searches. The incentives and programs change often.

Find Out What is Available in Your State, City, and Utility

If you have multiple locations, you need to compare your ability to save money based on government incentives. It is very easy to do that with their search tool. If you have a plant in California and another plant in Maine, the discounts and options available to you will be very different for the same cost reduction project.

For example, solar is heavily incentivized in California but hardly a factor in Maine. By understanding the contrast between the options available in each state, you can decide where to spend your money based on the best return.

If you can affect local policy, this is an exceptionally reliable database to use to build your case. If you or your local representative are trying to get incentives put in place for more renewable programs in your area, then you can research within this database for examples of what other government entities are offering.

DSIRE is a university-based program, and they are anxious to help you.

Use the Database to Calculate Solar Potential

There are solar calculators on the Internet, but DSIRE's is exceptionally easy to use. You enter your address, and the website will calculate what your potential savings. It then pulls the most up-to-date information about all the available incentive programs for your address. Nothing to figure out.

SUMMARY

I have used the DSIRE service for years, and I have found it to be remarkably comprehensive. The people that run it are very helpful. While your contractor may be informed about the incentives available in an area, it is not a given that they will know everything. It is incumbent upon *you*, the project owner, to flush out any additional savings that might be available to you through other incentives.

Many incentives go unused. The reason is it is a lot of trouble sometimes to apply for some of them. Many contractors who are installing systems for you may not go to the trouble to save you all the money that may be available. It is in your best interest to check on what is available yourself.

MINDSET

The Cost of Doing Nothing

We often assume all the risks lie only in future changes. There is a real and definable cost of staying where you are and doing nothing though.

We defend what we are currently doing to avoid the fear of an unknown future.

No one was ever fired for the energy project that didn't get done. Businesses and especially manufacturers often blindly favor the status quo over change. Nevertheless, these costs of staying still are real. Realizing that our inaction has far-reaching consequences is as important as considering the latest recommendation to change. To motivate us to act it is imperative that we take the time to understand our current Sword of Damocles.

Energy Savings Accumulate Every Month

Think of money going into a savings account at a bank. Energy cost reductions can be looked at in the same way—piling up every month. In the case of interruptible power and demand response, every month there are capacity payments that accrue. If there happen to be costs associated with making the change at any point, look at them like deductions from this ever-growing bank account of savings. When looked at from that perspective, the costs seem much smaller. When looked at from this "accounting perspective," even penalties the utility may impose are more palatable.

What if Money Is Not Available to Invest?

Often, energy is a manufacturer's most significant cost. A 10-15% reduction in those costs is like dropping money right to the bottom line of the business. The best part is those savings can usually begin quickly.

Why do manufacturing companies not take more definitive action? The fear and conservatism that grips most industrial facilities come from the pressure to maintain "uptime." Saving money requires a different mindset.

Having a small financial margin is stressful for the entire company. Focusing on energy costs problem can create a domino effect of savings in other areas the company and winch a financially stressed company out of its ditch.

This saving in energy can trigger a positive cycle of savings in countless other areas. This is something you receive automatically and organically by making others aware of energy savings as they occur. When we see our co-workers achieve something, we naturally want to compete. Seeing our success in our energy savings, our co-workers are also inspired to consider other energy sources. When people see things being accomplished by others, it triggers in them the belief that it *can* be done. Where they might have thought it impossible before, they now think maybe it can be done in their area too. Success begets success.

For example, as soon as Roger Bannister first broke the 4-minute mile, a number people ran sub-4-minute miles. Whether we believe we can or we believe we can't, it becomes a self-fulfilling prophecy.

Without proof that others have gone before us and gotten something done, we often subconsciously convince ourselves to stay where we are and do nothing. Once someone else accomplishes something, an indefinable force kicks in. We step out of our disbelief and rationalization. Expect results to magnify if you do a good job getting the word out about the energy savings you have achieved. The momentum will be contagious.

This phenomenon is seen in sales. You have probably heard the saying, *the best time to make a sale is right after you have made a sale.* Why is that? Because by succeeding, you are putting yourself in a frame of mind to take more risk. If you have just seen an example of taking an action payoff, and you are more inclined to do the same thing again because it feels good.

Until You Make Energy an Obsession, It Will Not Get Taken Seriously by Others

It is very easy to do things halfway, especially with energy. When we are not familiar with something, we are tentative. This is not much of a problem if you're goals are not that high, and you want to get credit for having done a project. But, it is incumbent upon you to fully commit to the accomplishment of energy cost reduction goals. The more fully you commit, the more you will look for alternate solutions. The more fully you commit, the less likely you are to give up when confronted with challenges and pushback.

There is a magic that happens when we are bold, and we commit. It is very difficult NOT to succeed when you put your whole heart and soul into something. It is very unlikely for any of our activities or goals to succeed without having to go through false steps and rejections. If you are not fully committed to success in an endeavor, you will not go the extra mile. You will not make it through the 99 failed attempts to get to the one victory. While it is possible that you could strike gold the first time you put your pick in the ground, it is a lot more likely that you will not and that you will need perseverance to find the gold vein.

It is not for the gifted few. Anyone can achieve results if they fully commit.

The power of commitment is something we all intuitively understand. Most people do not get there because it is a lot of work. The majority of people give up at the first or second attempt. The majority puts out 50% effort and expects 100% results.

Realizing that others have gone before you and accomplished these goals should give you the strength of purpose to push forward.

This concept is the reason I have been able to make a career out of energy consulting. You would not believe how many projects I have completed and how many millions of dollars I have saved for people who already knew the solutions but had given up on the way.

The reason I spent so much time on these mental parts of savings is that I have seen it alone derail many projects. Knowing HOW to do something is not nearly as important as deciding that you WILL do something.

Your chances of success are very high if you decide that you will not quit midstream.

Manufacturing companies are often filled with people who find it difficult to commit. They are much more comfortable with letting someone else determine their daily work schedule and their goals for the year. They are not likely to automatically buy in to your commitment, so do not expect them to.

When employees see lack of concern for costs in one area, they tend to ignore costs in other areas. As we have said, when you save money in one area, you get the free bonus of saving in other areas. Human nature and competition guarantee it. However, if you take no action and leave things the way they are, you not only do not benefit from the cost savings, but you will most likely generate more waste in your company.

How can that be? By not showing employees they need to be working together to reduce costs aggressively, you are teaching them that saving does not matter.

When money is tight, businesses usually turn to their banks as the best way to relieve the pain. More times than not, those same companies have an interest-free loan waiting for them hidden in the lines of their utility bills. Going through the process of getting a bank loan can be demoralizing. Months of proving that you don't need the money for the bank to shine their blessings upon you and give you the money. You should realize that you have a "bank" hiding inside utility charges that are waiting to be activated to give you a zero percent loan.

The great thing about seeking internal sources of funds is that the solutions are usually perpetual. I have seen companies achieve hundreds of thousands in refunds and then go on to save tens of thousands ongoing for years.

The interest you will pay to a bank is an extra burden on the business. That interest burden may go on for 30 years, and borrowing up to the limit can keep the business from being able to borrow money later to expand.

I encourage you to look at waste in all your payables and

especially your energy bills like a bank with no interest loans. You have at your fingertips a way to access money has zero interest anytime you want if you will roll up your sleeves. It is far easier to go to your own cost centers and reduce expenses than it is to go to a bank.

The devil is in the details. The more you focus on and understand the tiny elements of costs on a utility bill, the more you will save. And it is that attention to detail with any cost center that will drive success.

As you create a culture of cost reduction built on an examination of the details, you are showing everyone how to take responsibility for the business. It is very easy for people to sit on the sidelines in a manufacturing plant. If it is not part of their job description, many people have no interest. When employees work on the cost side of things they take more ownership.

Unwatched Energy Costs Will Drift Higher

Price increases by utilities can't be avoided. The only way to mitigate their impact is by aggressively cutting costs. Utilities and other suppliers have no obligation to let you know there are cheaper alternatives to the way you are currently being billed.

Why Do Utility Prices Always Go Up?

The power grid we use to distribute electricity has evolved into a behemoth. It is a "distributed generation" model whereby electricity is created in power plants and distributed over power lines. There are tremendous losses as the voltage is increased and decreased. It is expensive to grow and expensive to maintain.

In general, it is difficult to say what the future will hold with the current methodology. More efficient electricity generation plants like nuclear energy could proliferate under laxer environmental laws and the nodal model of delivery could continue to evolve. If this is so, prices will continue to rise as the aging energy delivery infrastructure continues to deteriorate and must be replaced.

Technological innovation could disrupt this trend by allowing for the economical production of power by individuals and companies.

As an example, Japanese companies have developed "pocket nuclear" units.

The one factor we have surprisingly little intelligence on is how quickly fossil fuels will deplete. Although we do know what our coal reserves are, we do not really know what oil reserves are worldwide. Large oilfields have been found to very quickly dissipate once they are close to the end of their lives. When geologists make predictions about how long oil fields will last, they don't understand how precipitous the final years of oilfield life might be. That means we could very quickly reach an endgame scenario.

While much beyond standard Li-ion is not available commercially, battery storage technology is moving forward at a tremendous rate. The advent of electric cars has created competition in this area that did not exist before. Solid state electrolytes, such as glass, offer energy densities and charging speeds ten times what we currently enjoy.

The Zen of Cost Reduction

Simply by watching something, you can influence it. The more you pay attention to something, the more you notice little details. The more you notice little details, the more you will are confronted and bothered by underlying injustices and inequities. You will want to fix them when you see them. Attention to the micro details, therefore, yields macro results across multiple categories.

When people know the numbers are being watched, they focus on improvements. When we feel like no one is paying attention, it is very easy to ignore the details. While you can benefit from having someone else look over your shoulder, you can also benefit by looking over your own shoulder by quantifying your daily actions. This may seem obvious to you, but very few people do it.

As an example, if you were to run around the track without timing yourself, you will, of course, finish in a certain amount of time. However, if there are three people in the stands looking down at you and watching you as you do it, you're going to run faster. If you were to time how long it takes to get around the track, your time to get around the track will decrease.

Once you make a note of the time something takes, you have a benchmark. Establishing benchmarks for whatever you do will naturally propel you to the next level. The mechanism behind this is not clear, but it works.

There is a movement afoot called the "quantified self." In a medical environment when people can monitor their own stats, their health improves. As we note those changes, then we can reflect on whether they mean anything or not. But as we "numerical-ize" the details we gain a measure of control.

There is no such thing as maintaining the status quo. We are always either getting better or getting worse at something. The sooner we realize this, the quicker we can improve manufacturing processes including energy costs.

It is a step forward if we decide we want to lower our energy costs. However, we need to start focusing on actual numbers. The quicker we start taking note of the energy numbers, the more we will think about them. The more we are thinking about it, the more constructive solutions we will come up with. Monitoring numbers always leads to progress.

FIVE ENERGY TRENDS

In general, the world will continue to consume more and more energy. It used to be that the United States was the leader but now other countries, including communist China, are adapting to lifestyles that demand a lot of electric power.

The following are the technological trends that will influence how cheap energy could get:

1. Cheaper Sensors

The Internet of Things concept describes the rollout of sensors and the connection of them through the Internet. Inexpensive sensor technology is allowing this to occur. The resulting data explosion is triggering the need to slice and dice the information.

If having information is a good thing and more information is a great thing, then we are moving into an information Nirvana. Just having the information will not be enough; our next struggle is knowing what to do with it.

As it relates to energy cost reduction, the information collected through an army of sensors deployed at the device level can give us real-time access to understanding areas of waste. Once inefficiencies are spotted, sensors can give the feedback needed to verify changes have been made to fix problems.

Recently, the Department of Energy held a competition among sub-meter manufacturers to develop a device level energy sensor for under $100. A prize was awarded, and vetting is currently being done.

2. Smart Meters

Online reading of meters is a trend that has been gaining speed for years because it cuts down on utility meter reading errors and eliminates the cost of deploying a meter reader. Being able to do it is expensive though, since meters must be replaced.

The advantage for utility customers, though, is enormous. Being able to access details of usage allows you to take advantage of power priced based on the time it is used. This can drastically lower costs.

Not only can you potentially purchase power cheaper but you can also drill down and understand patterns of consumption. These consumption trends can be compared with production schedules to identify areas of waste. Unfortunately, power companies have not been very quick to swap out dumb meters for smart meters.

This technology will also make usage data available on your smartphone. Along with bells and whistles to alert you of

approaching usage events like demand peaks, you will be reminded to take actions that will lower your energy costs. Knowing about the details triggers you to ask questions. Without the
data and the reminders that you are approaching critical peaks,
for example, life goes on as usual. Energy usage gets ignored,
and nothing changes. Ask your utility when they are going to
change to using smart meters.

3. Decentralization and Renewables

The power we require will be generated by us rather than
by large centralized utilities. In the past, people who would
have never gotten electricity any other way have been given
access to electricity through this centralized model. But this
model is outdated now. The high cost of transporting electrons
over long distances has offered entrepreneurs an opportunity to
decrease costs through localization of power generation.

Solar is just one example of this. Other ways to make power
locally include microgrids for neighborhoods. A group of houses
can go off the main power grid and power themselves directly
from a generator.

4. Electricity Replacing Gas

Electricity offers a far more predictable price. Natural gas
has disadvantages, not least of which is its toxicity. It is moved
across the country through underground pipelines and involves
risks.

Pipelines are buried only a few feet underground, and they
are not always well marked. It is therefore easy for road workers
to strike them with earth digging equipment.

Electricity, on the other hand, is much easier to control and
manage. While high voltage lines can be an eyesore, it is typically
transported hundreds of feet in the air when it is at high voltages. In some markets electricity is increasingly tied to the price of
natural gas. That is due largely to the fact that the fuel sources
for generating electricity are not as vulnerable to extreme weather
events as is the price of natural gas.

Hurricane season causes angst for all natural gas market-ers. Most offshore oil wells are in the Gulf of Mexico where the hurricanes occur frequently. Natural gas is usually found in con-junction with oil wells. Hurricanes can be disruptive and tend to put the off-shore oil wells temporarily out of commission. When this happens, the supply of natural gas starts dwindling and the price starts going up. It doesn't take long for the resulting super-high wholesale price to trickle down to you and me in the prices we pay at our homes and businesses.

Another driver of gas pricing is that it is an actively traded commodity and utilities are often under no compulsion to buy at the lowest price. The utilities that sell natural gas buy it based on market need and then pass the cost of the commodity part of the gas down on to their customers, homes, and business that inevitably feel the price increases.

When it comes to heating, however, natural gas has been, and still is, the fuel of choice. That is because it has been cheaper overall to purchase gas than it has been to purchase electricity. Heating is usually accomplished through an indirect flame of ig-nited natural gas. But it is challenging to control the temperature of such a directly applied flame. Processes that demand greater temperature control have always opted for electricity to achieve more precise adjustments.

5. Batteries

Batteries allow electricity to be used in remote locations or in times when grid-supplied electricity is not available. The abil-ity to store electricity cheaply in a small amount of space will be one of the most important topics going forward.

If we could store electricity more efficiencly now, most of the arguments against renewable energy would go away. Scal-ing the storing of electric power is a game changer. Renewable resources like solar, wind, and wave-generated power are avail-able free for the taking. While they cost nothing to harvest, they are intermittent and wreak havoc for energy users who would like to get off the grid but need a steady supply of power. But since their supply is intermittent, it is impossible to count on them to supply power when we want and need them to. For that assurance, we need better battery storage options.

Batteries are now primarily lithium-ion. The lithium-ion battery has some serious limitations, however. They must be charged very slowly because the electrolyte can catch on fire if charged too quickly or too much. The speed of charging aside, they are also notorious for developing a charging memory, which reduces useful life. A five-year-old lithium-ion battery will only be able to be charged to 80-90% of the maximum charging capacity it had when it was new.

However, there are solid-state battery technologies on the horizon could likely drastically change all this. In 2017, researchers instrumental in the development of the lithium-ion battery announced they had developed a battery with a glass electrolyte. The glass electrolyte is a much more uniform conductive media allowing much faster charging without the risk of overcharging or the electrolyte catching on fire. Charging times with this technology will decrease from hours to minutes and energy storage densities will be much higher.

Ultra-fast charging times for batteries with high energy densities—the amount of charge can be stored per volumetric space—will dramatically improve the market for electric cars. Now, even when hooked up to a fast charger at higher voltages it can still take hours to charge a lithium-ion battery in a smartphone fully.

When battery technology approaches the level of sophistication of power generation from fossil fuel sources, we will be able to store energy locally for long periods of time and then use it when we need it. That, in combination with an energy usage optimization strategy, will drastically change the landscape. That means it is unavoidable to think that we could create massive amounts of storage without simultaneously reducing the amount we consume.

As batteries evolve, technologies to take advantage of off-peak power and sources that are currently considered too inefficient now will emerge. Unfortunately, the success of lithium-ion batteries has slowed the development of advancing battery technology. As lithium-ion batteries have come down in price, it has not given tech startups in this space much incentive to expand research.

However, imagine a world where electricity storage was ubiquitous, not only at companies and homes, but on the electrical grid itself and at the generation plants. The whole electrical system would be converted into a virtual battery—manufacturing plants, homes, businesses, and even electric cars. Imagine after some time, a world where that energy was used more efficiently, where appliances were only turned on by AI when they were needed—not running on standby. In short, imagine a world of optimization controlled at the device level—a world of minimization of usage combined with maximization of storage.

That world is not that far away. Research is being done on the very system that will accomplish just that—artificial intelligence optimizing the entire system of storage and usage in a way that perpetually reduces the cost to produce and deliver energy to the consumer, whether that consumer is an industrial plant or a home, or even a self-driving automobile. We are only a few years away from this becoming a reality.

INERTIA OF THE STATUS QUO

Even when new technology solves our problems we tend to stick with the status quo. We wish for the new, and we say we want it, but the comfort of the old and inefficient pulls us way more than we would like to admit.

The comfort of the familiar is the reason new ideas of any kind take so long to take hold. Even obvious things that would improve people's lives that look like no-brainers to us now took eons to become mainstream. It is dangerous to be the technological torchbearer.

Those that wait for the technology to optimize and the price to come down may lose years waiting for mass economies of scale. The technologies to drastically reduce the energy costs in your plant are at hand. These changes will not be forced on you.

To take full advantage of these improvements you will have to be different than the masses who wait on social proof. This broad-scale superficial generalized "proof" that suffices for most of us in everyday life now will have to be replaced with deeper dives into the details of the technology.

Energy storage/batteries are really the golden bullets. If you can store enough energy onsite to power your entire facility without the participation of your utility for a few days, your options for cost savings go up astronomically. Not only does demand response become a no-brainer, but also buying energy real-time becomes easy because you can buy your energy cheap, store it, and use it later when you need it. It is the needing of electricity at certain times and not at other certain times that drives our costs now.

CO_2 SCIENCE AND THE BATTLE TO SAVE THE PLANET

As anyone who has an Internet connection or watches the 6 o'clock news knows (God forbid), scientists all over the world are batting the wiffle ball of universal truth back and forth trying to come to grips with the long-term and holistic effects of excess atmospheric carbon dioxide.

Global temperatures are rising slightly in certain places. Of that, there is little doubt. That can be blamed at least partially on the amount of atmospheric CO_2 more than what is utilized by living things. Burning fossil fuels (and breathing) creates carbon dioxide as a byproduct. CO_2 is a "greenhouse gas." Having too much of such a greenhouse gas in the atmosphere traps heat. It is believed too much CO_2 insulates the earth and makes temperatures rise and causes changes like the melting of polar ice caps.

When power plants burn coal to make power for factories, they expel excess carbon dioxide into the atmosphere—far more than can be absorbed by nature.

Most all utilities still make power by burning fossil fuels, and that typically means burning coal. The burning of coal releases carbon dioxide into the atmosphere. It is also believed that carbon dioxide must be "sequestered" if there is an excess of it above what the needs of plant life and photosynthesis are.

If you believe all this even a smidgen, the good news is by cutting your power demand and usage at your plant by even a little bit, you will coincidently be cutting down on the need for utility power generation facilities to make more energy and thereby put out more CO_2.

The jury is still out about the net effect of excess carbon dioxide in the atmosphere, but the consensus is that too much of it leads to adverse effects on crop production. First, the climate and the environment are complicated and intricately woven. With complex systems, it is difficult to predict broad impacts of small changes.

Ultimately, understanding the importance of CO_2 is about comparing the projected dangers that CO_2 present to the projected risks presented by other forms of power generation.

WHY EXCESS CO_2 MAKES TEMPERATURES RISE

Our planet is a miracle when you think about it. So many variables have conspired to make life work. If any of those myriad variables (temperature, gravitational pull, etc.) is off by a percent or two, the whole system will fall like a house of cards. (1) Physicist Geraint Lewis says it this way, "No matter which way we turn, the properties of our universe have finely tuned values that allow us to be here. Deviate ever so slightly from them, and the universe would be sterile—or it may never have existed at all."

One hundred million years ago solar energy was stored in

the form of organic compounds by the process of photosynthesis. When we burn "fossil fuels," that energy is released from carbon-based forms that store it, like coal and oil.

The atmosphere we breathe in one-way acts like the roof of a greenhouse, entrapping the gases that are created through this process. It acts to keep the earth warm and protects us from the utter cold of space ("The Problem of Carbon Dioxide," F. Niehaus IAEA Bulletin). The problem we are analyzing now, in fact, is not really about CO_2 itself, but rather the impact on the radiation balance in the atmosphere caused by relative amounts of these greenhouse gases.

2016 was Earth's hottest year. With an average land temperature of 58.69 degrees Fahrenheit, that is a full 1.69 degrees hotter than the average for the 20th century. That doesn't sound like much, but the difference in our current average temperature and the standard that we have discerned from climate records in ice and trees during the ice age is only 5 degrees.

From ice cores pulled in Antarctica, we can tell how greenhouse gas levels have affected our climate over time. Carbon dioxide is, in fact, the predominant factor in the greenhouse gas drama, followed by water vapor.

The planet and the atmosphere have changed a lot over time; this is not disputed. Atmospheric change, then, is the standard operating practice for our world. We should expect it. It is the rate of change that should make us pause and think.

Will our societies be able to adapt to a rapidly changing climate and the resulting lifestyle changes that those changes might impose? That is the question we face.

We don't know if 2018 will continue the trend toward a higher average temperature.

I would argue we have spent too much time splitting hairs over everyone's different predictions for the future and what excess CO_2 could mean. Like the stock market, we will never be able to sort that out before it is already upon us. What we do know is that global temperatures are rising faster than they have in recorded history. We know enough to have a gut sense that it matters on some level and we need to at least *slow down the additions of* CO_2 to our atmosphere until we can understand the overall impact of such changes better.

According to the intergovernmental panel on climate change in their report to the United Nations, it is 90% likely that accelerated warming in the last 50 to 60 years is due to these human contributions.

IS ENERGY CREATION THE ONLY CAUSE OF ATMOSPHERIC CO_2?

Not by a long shot, but it is the largest. Industrialization, large-scale deforestation, and the rise of combustion drive the excess CO_2 ever higher.

Combustion is a means of getting from here to there and producing energy—the energy we use to electrify our lifestyle and make the products we need and want.

As we have already discussed, the specific fallout of the increased temperatures is actively scientifically and politically debated, but what we think of as climate science had its origin in the mid-1800s.

Where has all this heat caused by CO_2 gone you may ask? A lot of it has made the oceans warmer, and at the poles, ice sheets have decreased *en masse*—Greenland, for example, lost 150-250 km cubic kilometers of ice per year. Between 2002 and 2006. The trapped greenhouse gases have nowhere else to go but to remain under the canopy of ozone.

WHAT HAPPENS IF WE DO NOTHING?

The answer is we don't know. But, what can be done to slow down the apparent tide until we get a better grip? How can we both have all the energy that we want and need and at the same time, cut back on CO_2 production?

GLOBAL EMISSIONS BY ECONOMIC SECTOR

Details about the sources included in these estimates can be found in the Contribution of Working Group III to the Fifth Assessment Report of the Intergovernmental Panel on Climate Change.

Energy Production—25% of 2010 global greenhouse gas emissions: The burning of coal, natural gas, and oil for electricity and heat is the most significant single source of global greenhouse gas emissions.

Industry—21% global greenhouse gas emissions (2010): Greenhouse gas emissions from production primarily involve fossil fuels burned on-site at facilities creating electricity. This sector also includes emissions from chemical, metallurgical, and mineral transformation processes not associated with energy consumption and emissions from waste management activities. (Note: Emissions from industrial electricity use are excluded and are, instead, covered in the Electricity and Heat Production sector.)

Agriculture and Forestry—24% of 2010 global greenhouse gas emissions): Greenhouse gas emissions from this sector come mostly from agriculture (cultivation of crops and livestock) and deforestation. This estimate does not include the CO_2 that ecosystems remove from the atmosphere by sequestering carbon in biomass, dead organic matter, and soils, which offset approximately 20% of emissions from this sector.

Transportation Industry—14% of 2010 global greenhouse gas emissions: Greenhouse gas emissions from this sector primarily involve fossil fuels burned for road, rail, air, and marine transportation. Almost all (95%) of the world's transportation energy comes from petroleum-based fuels, mainly gasoline and diesel.

Commercial Office Buildings—6% of 2010 global greenhouse gas emissions: Greenhouse gas emissions from this sector arise from onsite energy generation and burning fuels for heat in buildings or cooking in homes. (Note: Emissions from electricity use in buildings are excluded and are instead covered in the Electricity and Heat Production sector.)

Other things—10% of 2010 global greenhouse gas emissions: This source of greenhouse gas emissions refers to all emissions from the Energy sector which are not directly associated with electricity or heat production, such as fuel extraction, refining, processing, and transportation.

COUNTRIES THAT CONTRIBUTE TO CO_2 EMISSIONS

As of 2014, the worst offenders were China, the United States, the European Union, India, the Russian Federation, and Japan. These CO_2 emissions data are from all sources.

WHAT CAN WE DO ABOUT IT?

On a macro level, carbon sequestration is being pursued aggressively. That means that the exhaust stack on fossil fuel burning power plants must be scrubbed to remove the carbon dioxide to keep it from rising into the higher levels of the atmosphere. Once it is scrubbed from the exhaust gas, it can be "sequestered" or stored somewhere terrestrially—either under the ocean floor or increasingly, in a compressed gas state that may be able to be converted into an energy source itself at some point.

In a manufacturing plant, some things can be done to minimize the facility's impact on the environment.

**Actions You Can Take to Reduce the
Current Impact on the Environment:**
1. Measure your carbon footprint. By doing a greenhouse gas assessment, you can clarify what your actual contribution

to CO_2 emissions is. Once you start tracking this information, it can serve as a means of further reducing the amount of carbon that your operation is responsible for also known as your "carbon footprint." There are many free online resources you can use to estimate your carbon footprint.

2. Engage in carbon capping. This concept places a value on the amount of total carbon dioxide we are willing to emit into the atmosphere as a nation or as a collection of nations. When a value is established for a permit to emit carbon dioxide, then reductions in carbon dioxide at your facility result in an abundance over and above what you need that you have been permitted to email. That would be a sellable entity. You would be able to sell the carbon credit to another company because you have reduced your need for the total permit that limits the output at your plant.

3. Reducing energy demand and consumption. By reducing the amount of energy use at your plant, you are reducing the amount of energy that needs to be produced by your utility. In most areas of the country fossil fuel powered generation is still the order of the day. Therefore, if you reduce the amount of energy use in your facility you are causing your utility to have to burn less and thereby reduce the amount of carbon dioxide in the atmosphere.

Some other options that are doable at a plant-level are worthy of consideration. While they may be small steps relative to the major ones involving coal-based energy generation made at a societal level, they do contribute and encourage employees to get involved in a currently discussed related problem:

1. Rewarding employees for green commuting. Encouraging employees to not use energy-consuming equipment like cars to get to work each day. Green commuting can be a reality. Understanding your carbon footprint and thereby

promoting your company's desire to reduce its carbon foot print can have a significant impact not only on your carbon footprint but also on your employee's attitude toward saving money in other areas.

2.   Adding renewable energy sources at your facility. Both wind and solar energy are becoming more affordable. In some regions of the United States that have a high number of days with sunshine, it is becoming very popular to install solar collectors with paybacks that look like those of any other piece of equipment you might consider. This was not always the case, however. In the not too distant past, renewable energy projects were only for people willing to live with a very long payback. The prices have come down significantly since then, and even a little bit of renewable energy can offset your power bill significantly. While these energy sources are intermittent and not something you can count on 100% of the time, they can be excellent sources, especially if you let them charge batteries for later use.

By reducing your energy consumption through any of these means, you are contributing to the solution. You're setting an example for your employees which can have a ripple effect throughout the organization.

For our purposes, it is enough to recognize several more things:

1.   Climate change is important, but our generation will most likely never know how much so.

2.   You and your plant can make a positive impact by cutting demand.

3.   There is marketing value in reducing your carbon footprint.

It is clear that nothing here really offers the type of clarity we would like to have. With rising temperatures, some people will win, and some will lose. That means, in some locations increasing temperatures will enhance crop production and improve the livability of the place. Take for example, in a warming world, Siberia and Antarctica may become habitable. In other areas increasing temperatures are detrimental and may trigger a mass exodus. (Imagine Arizona even hotter.) For example, regarding health, warmer winters would mean fewer deaths, particularly among vulnerable groups like the aged. Other positive effects of climate change may include greener rainforests and enhanced plant growth in the Amazon, increased growth in the Amazon, increased vegetation in northern latitudes and possible increases in plankton biomass in the ocean. However, the main point may be as simple as keeping the current balance of power favorable to the United States. We have a vested interest in the power balance status quo. As Gregg Easterbrook writes in The Atlantic, "...when the global order already places America at No. 1, why would we want to run the risk of climate change that alters that order?"

Is it Even Possible to Understand Something with so Many Variables?

There is an avalanche of variables involved in diagnosing our climate ills. There are assumptions made, and factors are measured. There are also a number of factors that are not measured on purpose. Someone is making those choices based on some internal or external bias. There is a better than likely chance that we may have picked the wrong variables to analyze and the wrong benchmarks to use.

Science is iterative. It moves forward in fits and starts through trial and error. Here are the ways things used to work— Scientists suggest a hypothesis of how things work. Then they go about trying to disprove that hypothesis. The absence of disproof meant the hypothesis was valid.

Now, many climate hypotheses have a political dimension, and research is funded by people who are trying to prove a point. We must look at all this evidence with a skeptical eye.

How Do We Know Who to Believe?

It is hard to know who to listen to. Everyone seems to weigh in on the topic of climate science, frequently with a politically biased point of view under the guise of hard science.

Every side in the debate seems to be looking at a different set of data. Their conclusions are so opposite that we don't feel like we can believe any of them.

There is just too much to gain by maintaining the status quo, however. By stoking the coals of whatever side of the issue they happen to be on, politicians have been able to swing elections. Pandering to public biases and perceptions that each side has created for just such a purpose has become the way many get elected and reelected.

The power these issues offer to those willing to amplify the public's biases for one side or the other in the debate is worth more to leaders than getting at the truth.

The climate change debate has been devoured and regurgitated by politicians for their own purposes. It seems forever doomed to irrelevance for both sides of the discussion, while the politicians want to say we understand everything, climate science offers as many questions as answers. The rate of CO_2 expansion may be the problem, and you can help reduce the need for fossil fuel combustion by taking part in some current utility programs that will benefit your plant right now.

Index

Symbols
4CP 73

A
AI 184

B
bandwagon effect 17
batteries 182, 185
battery system 94
business interruption
        insurance 64

C
capacity and power rating 93
carbon dioxide 187, 190
climate change 192
concept level 35
CO_2 185, 186, 187, 188, 190, 191,
        194
coincident peak (CP) 73, 74, 75,
        76, 77
        reduction program 21
curtailment 22

D
demand 53
        response 45, 46, 64
                programs 37
DoD (depth of discharge) 93
deregulation 5
DSIRE 171, 172

E
economies of power
        distribution 128

economies of power
        generation 128
economies of scale 27
Electric Membership Cooperatives
        28
electric utilities 4
energy management system
        (EMS) 99, 97, 101, 105, 160

F
facilities charge 119
feedback loop 105
FERC Order 636 24
fossil fuel 13

G
greenhouse gas 185
grid networks 43

H
Hawthorne Effect 98, 99
hours use demand 49

I
interruptible power 37, 41, 42
interruptible programs 45, 138
interruptible service or "I.S." 35,
        38, 40, 45
        programs 44
        rates 41
interruption 42
investor-owned utilities 28
involuntary brown/blackouts 138

J
Joule effect 129

L
local distribution companies 23

M
maintenance manager syndrome
       14
Mayo, Elton 98
meter combinations 124
metering at higher voltage 127
multiple meters 126

N
net metering 89
non-compliance 36

P
peak sun-hour 86
pooled system 37
Public Service
       Commissions (PUC) 39
predominant use study 115
PV system 85

Q
quantified self 179

R
ratchet 117
rate analysis 142
real time pricing 67
rebills 119
relative risk 47
renewable certificates 91
risk 56, 57, 59, 61, 62, 65
       analysis 11
       assessment 149
round-trip efficiency 94
RTP 67, 69, 70, 71

S
sales tax 109, 110
       exemption 114
SCRECTrade 90
smart meters 180
solar PV system 85
solar power 79, 83, 86, 93
       credits 89
system lambda 69

T
totalizing 126
TOU rates 50

U
utility holdback 115

Printed and bound by CPI Group (UK) Ltd, Croydon, CR0 4YY

21/10/2024

01777086-0012